Published in 2017 by The Inamorata Press
18 Rosebery Road, Dursley, Gloucestershire GL11 4PT

Text © 2017 Matthew Watkins
Illustrations © 2017 Matt Tweed, Carol J. Watkins and Juliet Suzmeyan

ISBN 978-0-9564879-3-3

A CIP catalogue record for this book is available from the British Library.

Printed and bound by TJ International, Padstow, Cornwall, UK.

youareherebook.com

You Are Here

the biography of a moment

Matthew Watkins

"...they're burning candles now in Canterbury"

Martin Cockerham (Spirogyra, "A Canterbury Tale", 1972)

"There is no lovelier place in the world than Canterbury – that I say with hand on my heart as I sit in Florence – and I have seen Venice too."

Virginia Woolf (from a letter to her cousin Emma Vaughan, 1904)

"Canterbury... interesting place... unique, giant village in the Garden of England. You wouldn't believe it unless you'd lived here for forty years like I have how unique it is."

Richard Sinclair (from *Facelift* issue 10, 1993)

"People talk about the Cambry Pul theywl say any part of Inland you might be in youwl feal that pul to Cambry in the senter."

Russell Hoban (from *Riddley Walker*, 1980)

"...the streets ring out for every soul that thought and felt and passed through them in weakness and in strength"

Nicholas Heiney (from *The Silence at the Song's End*, 2011)

"In the first and final second of forever I thought of the long past that had led to now, and never... never... never..."

Robert Calvert (Hawkwind, "Ten Seconds of Forever", 1973)

table of contents

For all sentient beings

This work began its life as a local history book. Struggling to find a written history of the City of Canterbury that I could relate to, I'd decided to write one myself. Quickly, though, it mutated into something else altogether.

Flipping through the first few pages, you might think that it's a history of the Universe. It isn't. It's an attempt to describe the history of a moment. And to fully tell the story of that moment, I needed to give an account of both the history of Canterbury (where the moment "occurred") and the history of the Universe (in which Canterbury "occurred").

You'll find that this cosmic history quickly becomes Earth-centric, then Southeast-England-centric, human-centric (around the time humans start showing up in what's now the Canterbury area), then Canterbury-centric and, ultimately, author-centric. Just as this isn't a history of the Universe (despite starting out that way), it's not an autobiography either (despite sort of ending that way). But in order to tell the story of a moment as fully as possible, the life of the person experiencing that moment inevitably got involved.

And, as the chosen moment approached, the telling of the story itself also became part of the story. Some might think of this as postmodern. If it is, it wasn't planned or intended that way, it just happened as an unavoidable result of the process I set in motion for myself.

Anyone attempting to describe the history of a moment would reveal something of what they (or their culture) prioritised and how they understood time. A medieval monk or Amazonian tribesperson, charged with this same task, would produce radically different descriptions. But I'm me, and I'm here and now in this confused, fragmented yet hyper-connected post-everything Western culture, so this is how it came out:

Cantwaraburgh, July 2017

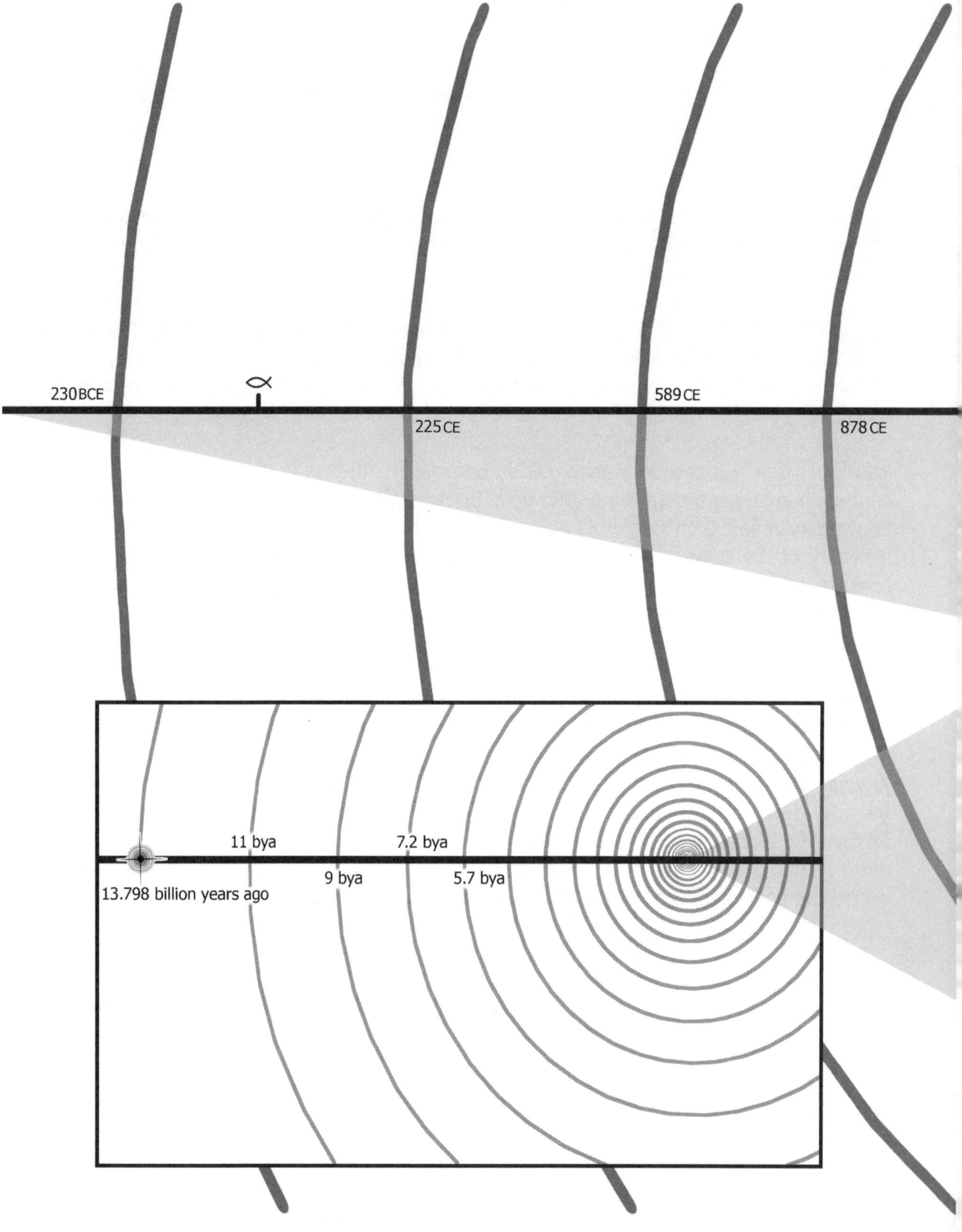

230 BCE

225 CE

589 CE

878 CE

11 bya

9 bya

7.2 bya

5.7 bya

13.798 billion years ago

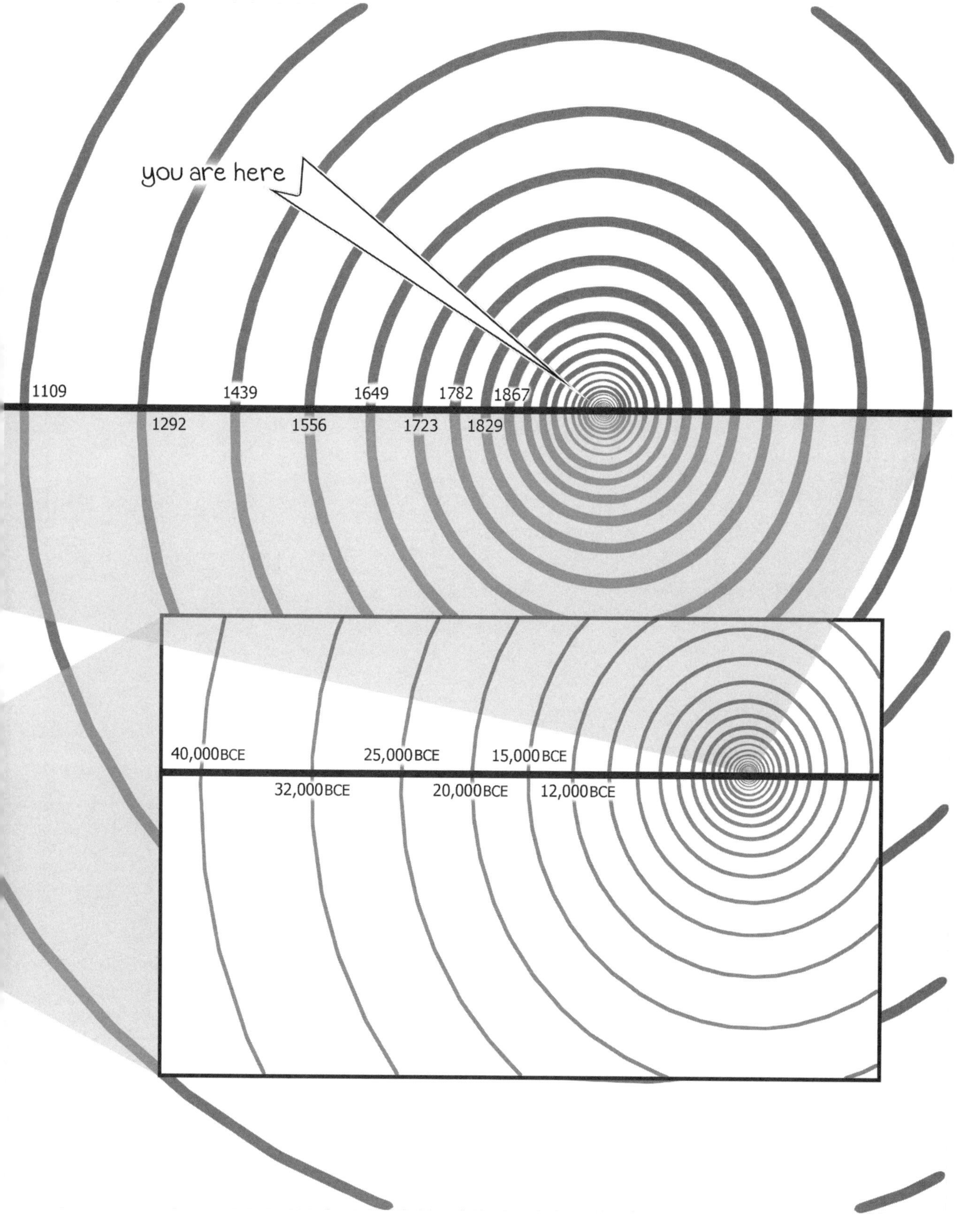

you are here

1109

1292

1439

1556

1649

1723

1782   1867

1829

40,000 BCE

32,000 BCE

25,000 BCE

20,000 BCE

15,000 BCE

12,000 BCE

13,798,000,000 to 11,000,000,000 years ago

BIG BANG. That's what they call it. It may have come to seem ridiculous by the time some people read this, but it's the story I grew up with. This Universe, I've been told, began 13.798 billion years ago, give or take a few million years. From out of nowhere, in the midst of nothingness, an explosion occurred, bringing forth all the ingredients of the early Universe.

Matter and energy were initially merged in a way that's hard to make any practical sense of but they soon separated out, material particles then starting to cluster together to form nuclei of atoms, then full atoms.

Although everything was flying away from everything else as a result of the initial explosion, after a few hundred million years gravity was causing large enough quantities of hydrogen atoms to clump together under sufficient pressure to ignite into stars. These stars then began to clump together into galaxies and these galaxies into galactic clusters. Heavier atoms were being forged in the interiors of stars, these being spewed out in supernova explosions, ingredients for future planets.

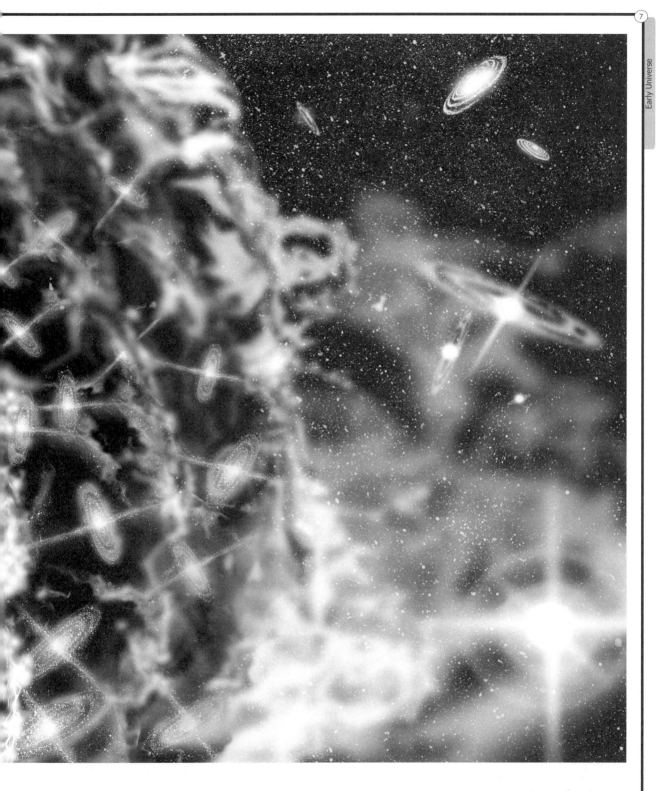

The maverick psychedelic philosopher Terence McKenna gleefully described the Big Bang Theory as *"The limit case of credulity: that the entire universe sprang from nothingness, and at a single point and for no discernible reason"*, quite reasonably adding that *"if you believe this, you can believe anything!"*

11,000,000,000 to 9,000,000,000 years ago

Stars and galaxies continued to form. The beginnings of the galaxy we've come to know as the Milky Way were already in place, its "galactic disc" forming in this period. Its nearest neighbour, the Andromeda galaxy, had also taken shape and galaxies elsewhere were merging, forming clusters and superclusters. Also being created across the Universe were globular star clusters, quasars, pulsars and even some planetary systems.

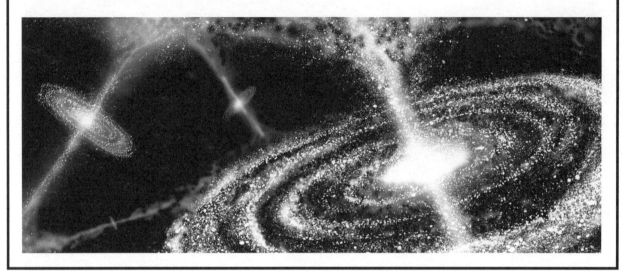

9,000,000,000 to 7,200,000,000 years ago

By this stage, galaxies were beginning to look rather like how they do now. Star formation seems to have been at its peak and the spiral arms of the Milky Way were formed around this time. The material content of the Universe was now organising into larger structures – filaments and sheets consisting of galactic clusters and superclusters interspersed with void spaces.

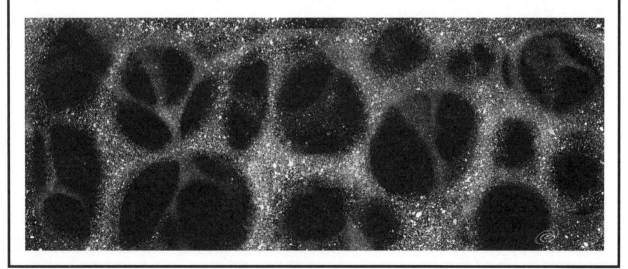

7,200,000,000 to 5,700,000,000 years ago

According to recent (1990s) cosmological theory, shifting relations between matter, radiation and the underlying vacuum around this time meant that the "matter-dominated era" which had been causing the gradual slowing of the expanding Universe gave way to a "dark-energy-dominated era" in which the expansion began to accelerate.

5,700,000,000 to 4,600,000,000 years ago

The system involving the stars we now call Alpha Centauri A and B had formed, out towards the edge of the Milky Way.

4,600,000,000 to 3,600,000,000 years ago

Relatively near Alpha Centauri (in galactic terms), a giant interstellar cloud had collapsed down to create that most familiar of stars, the Sun, igniting as an enormous hydrogen-based nuclear reactor. From the cloud of dust and rock around it gradually coalesced several orbiting planets, including the Earth. The early Earth wasn't as large as it is now, its diameter having considerably increased by hundreds of millions of years of bombardment by foreign objects. It's now widely believed that one near collision with a very large object led to a chunk of the (still very unstable) Earth being pulled away, this then forming into the Moon.

A combination of these collisions and radioactivity heated the Earth's interior, causing it to separate into a liquid metal core and a molten "mantle". Eventually a brittle rock crust formed on the surface but this was being continually broken up and reformed by volcanic activity from below and collisions from above. The atmosphere settled into a mixture of about 70% water vapour plus hydrogen sulphide, methane, ammonia and carbon dioxide, resulting in global greenhouse conditions. When the temperature finally fell below 100°C, the planet was subjected to torrential rain, thus beginning the water cycle and the formation of oceans.

3,600,000,000 to 2,900,000,000 years ago

From a hundreds-of-millions-of-years-old soup of self-replicating molecules, two of the three currently recognised branches of life, archaea and bacteria, appeared. Their cells then rapidly spread across the planet, which was almost entirely underwater at this point.

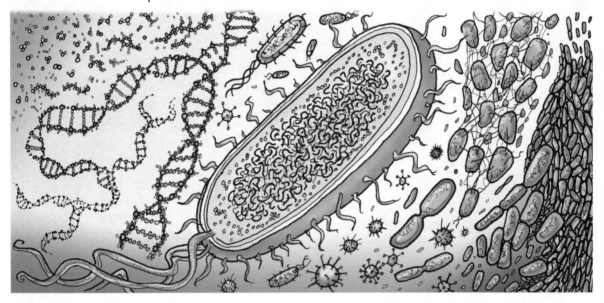

2,900,000,000 to 2,300,000,000 years ago

A permanent terrestrial crust formed over the whole planet, a mosaic of plates drifting around atop the mantle, their collisions forming mountains. Although earlier landmasses (given names like "Ur" and "Vaalbara") may have existed, the first continental core we can be reasonably sure about, Kenorland, rose from the sea around this time.

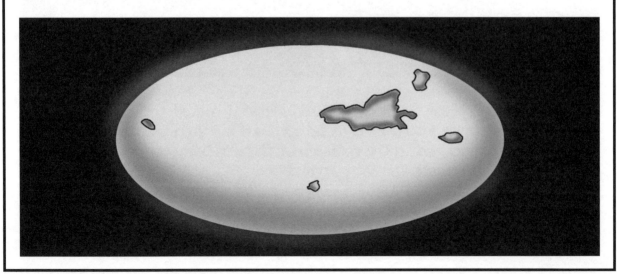

2,300,000,000 to 1,800,000,000 years ago

The first examples of the third branch of life, eukaryotes, now appeared. These were a new type of cellular life – amoeba-like "protists", the forerunners of multicellular plants, animals and fungi.

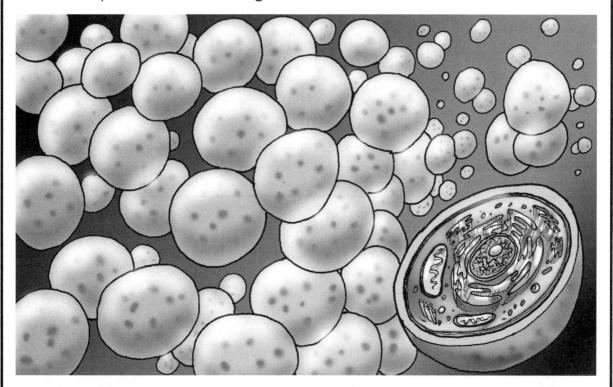

1,800,000,000 to 1,500,000,000 years ago

Kenorland had broken up by now and from its fragments the supercontinent Columbia formed. It encompassed almost all of the planet's landmass before breaking up.

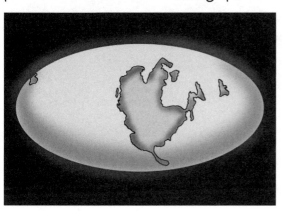

1,500,000,000 to 1,200,000,000 years ago

As primitive life continued to evolve in the sea, bits of Columbia drifted around the planet for hundreds of millions of years.

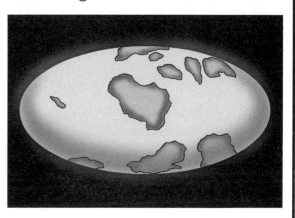

1,200,000,000 to 950,000,000 years ago

Some of these bits eventually collided to form the supercontinent Rodinia.

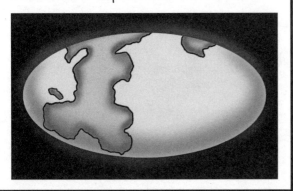

950,000,000 to 750,000,000 years ago

Rodinia started to break up.

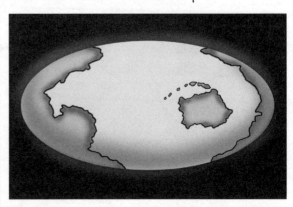

750,000,000 to 600,000,000 years ago

The first multicellular lifeforms, chains of cells, began to appear.

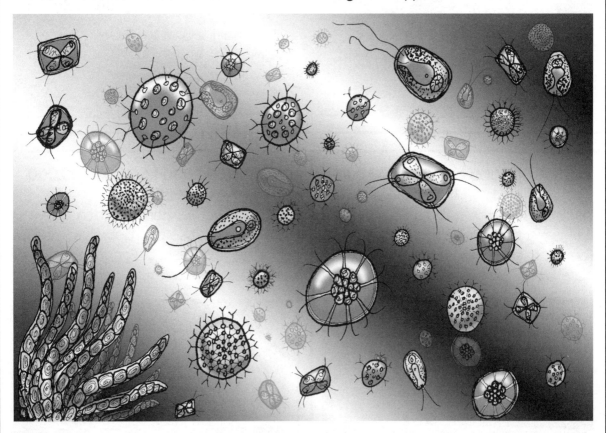

While Rodinia was breaking up and reforming into the supercontinent Pannotia some minor volcanic activity began to create a small blob of land that was to become the microcontinent Avalonia.

600,000,000 to 475,000,000 years ago

Avalonia had now formed as a landmass. It was in the Southern Hemisphere near the edge of a new supercontinent, Gondwana, but soon began to drift off to the north.

The "Cambrian explosion"! A sudden proliferation in biodiversity! Trilobites everywhere!

A series of ice ages then caused mass extinction, wiping out 95% of all life. After the last of these, the warming that set in encouraged a whole new wave of undersea fauna including sponges, jellyfish and horny-shelled brachiopods.

475,000,000 to 375,000,000 years ago

Gondwana drifted to the South Pole and started to freeze, causing more extinctions, while Avalonia drifted up to collide with the Laurentia and Baltica plates and become part of the northern continent Proto-Laurasia.

While the first vertebrates were evolving in the ocean and three-metre-long sea scorpions ruled the seabed, plants were making their way out onto the land, moving from coastal regions, swamps and marshes, eventually forming simple forest ecosystems involving horsetails and lycopods (proto-trees a few metres high).

375,000,000 to 300,000,000 years ago

Gondwana drifted north again and thawed, eventually joining up with Laurasia to create mega-continent Pangaea, Avalonia near its centre. Driver, follow that microcontinent!

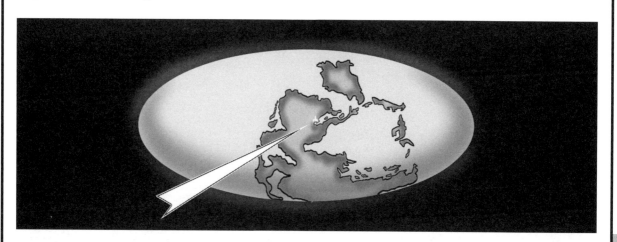

Amphibious arthropods were venturing onto the land, along with spiders, centipedes, and crustaceans, the latter then evolving into wingless insects. The amphibians (up to three metres long) evolved into early reptiles, then egg-laying tetrapods who were able to leave the coasts and conquer inland regions. Swamp forests now included plants up to fifty metres high.

A series of meteor strikes and volcanoes threw up huge amounts of dust, blocking the sun's rays. Global temperatures fell, icecaps grew, sea levels dropped, many species became extinct and vast rainforests died (a source of much future coal, this being the "Carboniferous Era").

300,000,000 to 250,000,000 years ago

There was now one landmass, Pangaea, and a vast ocean. Pangaea's size meant that its interior received almost no precipitation and so was mostly desert. From the existing spore-based plant species, conifers and ginkgos (seed-based gymnosperms) evolved to deal with the lack of water and spread inland. Meanwhile, reptiles and insects were getting around the problem of water dependence by developing hard-shelled eggs and pupae.

Reptilian land animals called therapsids evolved a sub-class called cynodonts which displayed some early mammal-like features in their bones, locomotion and metabolism. Driver, follow those species! The first dwarf dinosaurs appeared too, and some flying reptiles, but volcanic eruptions on the future Siberian landmass then caused global greenhouse conditions, making 90% of species extinct.

250,000,000 to 200,000,000 years ago

During this warm, dry period ferns and horsetails conquered the desert while archosaurs (running false crocodiles) were the dominant land animal. By the end of this period dinosaurs had evolved from archosaurs (some already taking to the air) and the archosaurs had become extinct.

200,000,000 to 150,000,000 years ago

Pangaea was breaking up in this time of extreme heat and high humidity. A massive earthquake caused a tectonic gap which was eventually to become the Atlantic Ocean.

Although dinosaurs continued to dominate among land animals and conifers were doing very well, by this time arguably the first true mammals (small furry ones) *and* the first angiosperms (flowering plants, starting with species like magnolia trees) had emerged.

150,000,000 to 125,000,000 years ago

Pangaea had split back into Gondwana (south) and Laurentia (north). Avalonia was now part of Laurentia but had been split into eastern and western parts. The western part was to become part of the east coast of Canada, the eastern part ending up on the western fringe of Europe. Gondwana started to break up into what would become South America, Africa, Australia and Antarctica.

Dinosaurs continued evolving, the first primitive birds appeared (feather-clad dinosaurs) and, despite conifers thriving in the form of cedars and redwoods, flowering plants and trees were effectively taking over, pollinator insects co-evolving with them. Monotremes (small furry egg-laying proto-mammals) were evolving from cynodonts.

125,000,000 to 100,000,000 years ago

A great sea covered the eastern part of Avalonia, sedimentary deposits gradually accumulating on its bed which were to become a type of stiff blue clay now known as "the Gault". Marine lifeforms present would have included crabs, fish, bivalves, gastropods, ammonites and squid-like belemnites.

100,000,000 to 75,000,000 years ago

Sea levels rose, as did temperatures, and as less and less sediment entered the sea, foraminifera (a type of amoeboid creature) thrived in its clear, warm waters. For thirty-five million years their calcium-rich skeletons fell to the clay seabed creating a thick layer that would eventually become the chalk downlands of southern England.

Among the mammals, a class of proto-primates evolved. These were small nocturnal insectivores resembling rats.

75,000,000 to 60,000,000 years ago

A ten kilometre meteorite slammed into the planet, darkening the sky for millennia and setting off tsunamis which devastated coastal habitats, as well as earthquakes and volcanoes. 70% of species became extinct, including almost all of the dinosaurs. The exceptions were birds and crocodiles, these making it through along with some of the mammals and flowering plants, both of which forms of life had been thriving before the catastrophe.

60,000,000 to 50,000,000 years ago

Enormous flightless birds dominated the plains, although small songbirds also emerged at this time. Tropical forests were thriving (the future source of all the lignite in Britain and elsewhere) and mammals kept diversifying. As well as "horned giant" rhino ancestors, many smaller mammals also developed. Rodents, bats, cloven-hooved animals, odd-toed ungulates and sloths appeared, as did the first true primates.

50,000,000 to 40,000,000 years ago

After a transitional amphibious period, some species of mammals returned to the oceans, becoming the first true marine mammals (ancestors of whales and dolphins). On land, the earliest horses evolved.

40,000,000 to 30,000,000 years ago

Antarctica drifted to the South Pole and iced over, while Australia and South America were finally isolated. Many primitive mammal lines died out as the climate became cooler and drier (a couple of meteorite impacts at this time may also have contributed). The interiors of continents were taken over by scrubland and semi-desert.

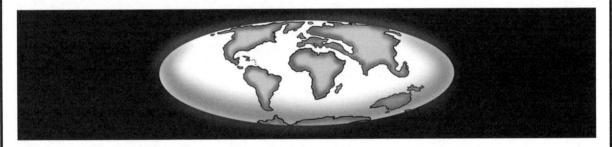

Huge, almost dinosaur-sized mammals developed, as well as rhinos, camels, horses, rabbits, pigs, and carnivores. Carnivores had soon split into feliforms (cats, mongooses, hyaenas) and caniforms (dogs, bears, wolves, otters, seals). Certain primates had the largest brains among all the land mammals from this point on.

30,000,000 to 25,000,000 years ago

Both on the northern continents and in the sea, animal life was starting to resemble a lot of what we're familiar with now. Flowering plants and trees continued with their unstoppable rise, temperate deciduous forests starting to replace tropical coniferous ones.

As the continents kept drifting towards their current positions, tectonic plate pressures continued to shape mountain ranges in North America (the Rockies), Asia (the Himalayas) and Europe (the Alps).

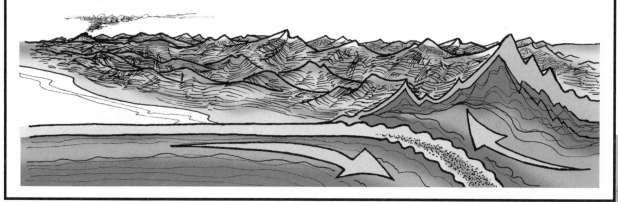

25,000,000 to 20,000,000 years ago

The world's landmasses were beginning to be conquered by grasses. Herds of grazing animals emerged to exploit this new food resource, predators then evolving to prey on them. In the primate world, hominids (great apes) had just broken away from Old World monkeys.

20,000,000 to 15,000,000 years ago

The *Hominidae* family of primates diverged from the *Hylobatidae* family (gibbons).

15,000,000 to 12,500,000 years ago

The tectonic plate movements which were continuing to shape the Alps were also lifting up a chalk-covered corner of eastern Avalonia, an area one day to be known as the Weald. African great apes (*Homininae*) evolved from *Hominidae*.

12,500,000 to 10,000,000 years ago

Grasslands had established themselves around the world: savannahs in Africa, prairies in North America, pampas in South America. Ants and termites, snakes and grassland mammals were beginning to diversify. Larger horses were also appearing.

10,000,000 to 8,000,000 years ago

The *Hominini* tribe, consisting of proto-humans and chimpanzees, split from the *Gorillini* tribe, consisting of gorillas. (A tribe is larger than a genus, but smaller than a family.)

8,000,000 to 6,500,000 years ago

The subtribe *Hominina* (humans and their biped ancestors) split from the subtribe *Panina* (chimpanzees), came down from the trees and adapted to a new kind of life on the plains.

6,500,000 to 5,000,000 years ago

By now, the Weald had some rivers running off it north and south. One of these, flowing northwards, had resulted from a series of linked fault lines that had weakened the chalk, allowing springs and streams to emerge and thereby creating a meandering line of erosion. Millions of years later this was to become known as the Great Stour Valley.

5,000,000 to 4,000,000 years ago

Mammoths make their first appearance in the fossil record.

4,000,000 to 3,200,000 years ago

The *Australopithecus* genus of proto-humans evolved in eastern Africa, among the first bipedal primates. They were mostly gathering nuts, berries and roots, and using simple tools.

The first modern lions, rhinos, zebras, giraffes, elephants and gazelles also appeared around this time.

3,200,000 to 2,500,000 years ago

The current ("quaternary") ice age began. It was to be a series of slow glaciations interspersed with warmer periods.

2,500,000 to 2,000,000 years ago

The first species of the genus *Homo* show up in the fossil record. By the end of this period they were using axes, hunting in groups and starting to lose their furry coats.

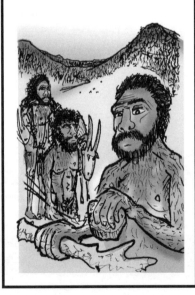

2,000,000 to 1,600,000 years ago

The last of the species of australopithecines had by now become extinct. *Homo erectus*, a new hominid species, emerged and started moving into Eurasia. Humans were beginning to make more sophisticated stone tools and building shelters.

1,600,000 to 1,300,000 years ago

Homo erectus continued moving into Eurasia and (at least in Africa) were practicing the controlled use of fire. Their brain size continued to increase.

1,300,000 to 1,000,000 years ago

Homo antecessor evolved, one of the first (now extinct) human species to venture into Europe.

1,000,000 to 800,000 years ago

Flint tools discovered on the coast of eastern Avalonia (later to be known as Britain) have revealed that hominids, probably *Homo antecessor*, were inhabiting such cool northerly climes (something like Sweden is now), seemingly without clothes or fire. They'd have been living largely on the meat of elephants, rhinos, elk, bears and lions, edible plants being scarce. These humans seem to have mostly inhabited marshes and flood plains along the course of the River Thames, as it was then.

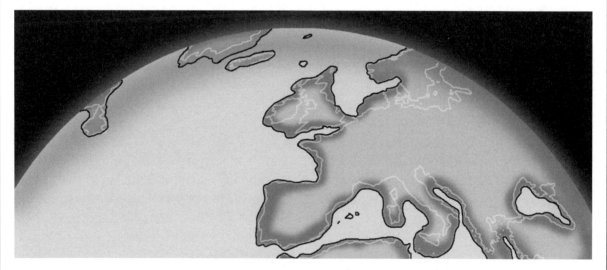

Britain was still part of continental Europe at this time and to have reached it these people would had travelled across a 200 km wide stretch of land. This would have involved crossing the future Great Stour Valley, on the border between a band of 100-million-year-old chalk and sedimentary deposits of sand and clay from shallow seas which were present forty to fifty million years later.

800,000 to 650,000 years ago

Homo erectus may have been in Europe at this time.

650,000 to 520,000 years ago

Homo heide!bergensis had evolved, a species of early humans who eventually lived in Africa, western Asia and Europe.

520,000 to 400,000 years ago

Teeth, bones and flint tools of *Homo heide!bergensis* have been found in southern Britain from this time. Elephants, wild horses and rhinos were roaming the land, being hunted and butchered with flint implements. Humans were involved in organised hunting – thinking, planning and working in groups.

The Anglian glaciation occurred around this time, the most extreme of the last two million years and the one in which the ice sheet came the closest to the area of the future East Kent that we have our eye on. During these glaciations, Britain would have been inhospitably cold so human populations would have gradually receded back onto the continent as the ice crept in from the north.

400,000 to 330,000 years ago

Neanderthals and *homo sapiens* diverged from a common ancestor, probably either *Homo heidelbergensis* or *Homo antecessor*. Both of these new species of humans appear to have had the rudiments of language.

330,000 to 250,000 years ago

A warm interglacial period gave way to another wave of glaciation (the Saale or, in Britain, Wolstonian glaciation). As well as clothing, burial rites (arguably the beginning of religion) had emerged in Neanderthal social groups, and Neanderthals had now started venturing into Europe.

250,000 to 200,000 years ago

Modern humans (*Homo sapiens sapiens*) first appeared in Africa, anatomically indistinguishable from yourself. A couple of brief, warmish spells occurred in Britain.

200,000 to 165,000 years ago

Much of Britain was now under an ice sheet, as far south as the Thames Estuary. But Neanderthals were not far away, in what's now France. Living in isolated groups and not venturing far from their home bases, their diet included woolly rhinos, mammoths and reindeer. They survived the glacial periods in Europe and had the continent to themselves for over 100,000 years.

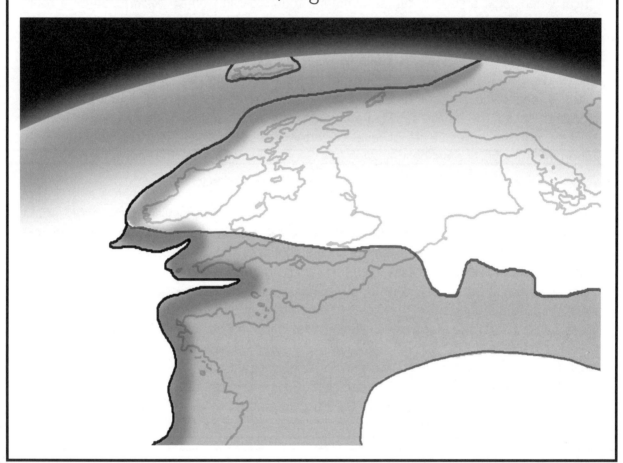

165,000 to 130,000 years ago

The Wolstonian glaciation receded.

130,000 to 100,000 years ago

The Eemian (or Ipswichian) warm interglacial period occurred during this time. It was the second to last, somewhat warmer than the one we're in now (the Holocene). Elephants and hippos were roaming Britain again, as in previous warm spells. Neanderthals were definitely in southern Britain by now and working with increasingly sophisticated stone tools.

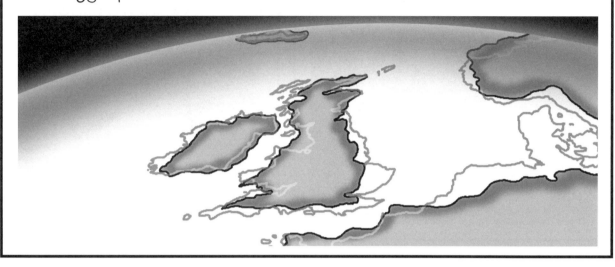

100,000 to 80,000 BCE

Some groups of humans had started to adorn themselves with ornamental beads, etc. and the nomadic tribal hunter-gatherer lifestyle had by now emerged in various places. The Sahara was wet and fertile at this time.

80,000 to 65,000 BCE

The tusk of a woolly mammoth from 75,000 BCE has been found near Swalecliffe, just north of Canterbury on the Kent coast. The Devensian glaciation (the last one) would have been underway at the time. The ice sheet wasn't to get further south than Norfolk but the glacial period would last until about 12,000 years ago.

65,000 to 52,000 BCE

A human population explosion began across Europe and Asia. Cave burials were going on, with the emergence of a kind of ancestral cultism and the use of ritual body paint.

52,000 to 40,000 BCE

Britain was still largely under an ice sheet but there was some *homo sapiens* activity, at least in the southwest (the earliest material evidence for this coming from Kents Cavern near Torquay).

40,000 to 32,000 BCE

Homo sapiens had become dominant in Europe, with "upper palaeolithic" tools: slender flint blades with handles and spear shafts. They were also using tools carved from bone, ivory and antler. Their culture included cave art, jewellery, musical instruments and grave goods, and there's evidence of extensive social networks and complex language.

32,000 to 25,000 BCE

Homo sapiens started to turn up in Britain (once again joined to the mainland) in larger numbers. Neanderthals gradually started to disappear after a few millennia of co-existence — it's still not clear quite why or how.

25,000 to 20,000 BCE

The Devensian glaciation reached its peak during this time. Southern Britain was now a treeless tundra and humans appear to have been driven out altogether. Many small fertility goddess figurines from this period have been unearthed.

20,000 to 15,000 BCE

Most of Britain was now under ice almost a mile thick but the sheet didn't extend into the southeast. Although mammoths had largely disappeared elsewhere in Northwest Europe they were still hanging on in that corner of Britain. Elsewhere in Europe, humans were developing ever more sophisticated tools, weapons and musical instruments.

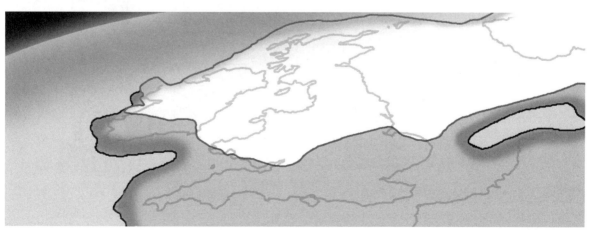

15,000 to 12,000 BCE

Temperatures started to rise again. Humans have continuously occupied Britain from this time on. Lions, rhinos and elephants were soon wiped out and the domestication of dogs (from the descendants of wolves) began.

12,000 to 9000 BCE

Although this period still saw considerable fluctuations, temperatures in Britain started to approach current levels. Red deer and wild horses became the dominant native mammals, while tundra-dwelling mammals like mammoths disappeared. Birch trees, shrubs and grasses started to spring up in the thawing tundra.

9000 to 6750 BCE

Melting ice sheets led to higher sea levels and Britain again becoming an island. Tundra and open steppe were gradually being replaced by forest – birch and pine, then hazel, elm, oak and lime. The coastline was now more or less as it currently is and the Great Stour had found a new eastward course to the sea.

The human inhabitants lived as hunter-gatherers, eating wild plants, fishing and hunting red deer, roe deer and wild pig. New types of sophisticated arrowheads were developed, as well as drill bits for making stone beads. Flint blades were used for harvesting and glues were used to hold blades into handles. Elk antler mattocks were used for digging roots, etc.

6750 to 5000 BCE

"Mesolithic" culture was now well under way in Britain, with stone chambered tombs being built. A type of shamanic religion appears to have been practiced. Houses were starting to be built as dwellings, bows and arrows were being used and forests were sometimes burned to attract grazing animals. The environment was increasingly wooded and human activity tended to be focussed along river valleys. Mesolithic stone tools have been found in Perry Woods near Canterbury, evidence for human activity in the Stour Valley at this time.

5000 to 3550 BCE

Social complexity was increasing in Britain, with a tendency towards settlement. A few examples of semipermanent coastal villages have been found with food stores, shell middens (dumps), territorial boundaries and cemeteries but there was more of this in Europe than in Britain. Britain tended to have smaller, temporary settlements without obvious storage facilities. These would have been seasonally occupied by people moving up and down major rivers. Land mammals were the main source of protein but there seems to have been a shift towards exploiting coastal seafood resources by the end of this period. There's no evidence that the oyster beds on the East Kent coast (where Whitstable now is) were exploited at this time, though.

The Neolithic revolution, involving the farming of crops and animals, began. A new wave of human settlers appears to have arrived in Britain with axes, seeds, domesticated cattle and sheep. They cultivated barley, made pottery and seem to have coexisted with the established populations. The landscape came to involve a **network of isolated farmsteads** but not yet villages. Not everyone settled down and started farming for some time – some groups of humans continuing to live as hunter-gatherers.

3550 to 2420 BCE

A polished jadeite axehead made from stone quarried in northern Italy was left in the future Canterbury area around this time. This would have been an extremely valuable ritual object (it's now in the British Museum) and is further evidence for both human activity in the area and a wide geographical network of connections. Stone bowls and other artefacts have been found in the area from this time but the jadeite axehead remains the most spectacular find. The existence of ritual axes suggests that there was an interplay between the spiritual and practical aspects of life. Pottery decoration and monumental architecture styles were similar across large areas, suggesting that practice something like pilgrimage was going on.

Not far upcountry, megaliths such as Kit's Coty, Little Kit's Coty and the Coldrum Stones were being constructed in the Medway Valley. Further west, the great monuments of Avebury and Stonehenge were taking shape. Closer to home, a Neolithic-style long barrow now known as "Juliberrie's Grave" was constructed beside the Stour near what would eventually become the village of Chilham.

Juliberrie's Grave seen from across the Stour Valley in the 1720s

Metalwork first started to appear towards the end of this period, marking the close of the Neolithic era. Increasing amounts of land were being cleared for farming and the first villages began to appear.

2420 to 1520 BCE

The Bronze Age had begun. More forests were cleared and stone circle building began in parts of the country. Neolithic long barrows were giving way to round "tumuli" across Britain, some containing elaborate grave goods. The uniformity of "beaker pottery" across Western and Central Europe during this time has led to assumptions about a migrating "beaker people". But by 2400BCE copper mining was under way and bronze was circulating so it may be that with the spread of these raw materials there was a spread of behaviour patterns, artistic styles, ideas and beliefs.

Bronze items were being made with a fired wax-and-clay system: tools, weapons, shields and cauldrons. A lot of the bronze artefacts that have been found seem to have been intentionally thrown into bodies of water for ritual purposes or else buried in hoards. Gold jewellery from this time has been found in East Kent.

Barley, wheat, cattle, sheep and pigs were being farmed, initially on a small enough scale that lasting marks weren't made on the landscape. Eventually communities began to divide up and enclose land on a much wider scale, traces of this still being visible in parts of southeast Britain. Particularly on the southern chalklands, an awareness of fertility the need to nurture soil had arisen – there's evidence of manuring practices from that time. Grain storage pits were starting to become widespread and roundhouse villages were appearing.

1520 to 800 BCE

By now there was a lot of cultural and economic exchange between people in Portugal, Spain, France, Britain and Ireland, with further trading links into Scandinavia and the Mediterranean. As a result, these peoples began to display cultural resemblances and have come to be known as "Celts", a cultural rather than racial grouping, possibly with a unifying language. Some historians think Celtic culture came later from central Europe but, in any case, inhabitants of the Stour Valley in southeast Britain would have seen a lot of traders passing through, mainly across from Northern France but occasionally from far-flung locations on the Atlantic seaboard and elsewhere.

Spelt, an early form of wheat, may have been introduced into Kent around this time. Another import was a new burial practice: cremation, with urns buried in cemeteries.

800 to 230 BCE

The Bronze Age ended with the coming of iron. Peoples in the British Isles and Western Europe were now speaking Celtic languages. There's no real evidence for a "Celtic immigration" into Britain as once believed – it seems to have been more a flow of culture. This would have been driven by the expansion of trade on the western seaboard. Settlements in the southeast of Britain were larger than those in the southwest, involving some substantial villages as well as isolated farmsteads. Crops were being grown and animals reared in an integrated arrangement involving feeds and manuring, while hunting was on the decline. Salt from the Kent coast at Seasalter was being produced and probably traded, while metal goods were coming over from the European mainland, including bronze items in the "La Tene" art style.

From around 350BCE, a hillfort now known as Bigbury was occupied just above the Great Stour Valley. Around the same time, a couple of miles away, the oldest currently known human settlement in what was to become Canterbury was established (near the corner of Castle Street and St. John's Lane). This was a small fortified settlement near a fordable section of the Stour. The relationship between the people who built this and the people up at the hillfort is not clear but with the coming of iron axes, people were starting to clear undergrowth and dwell in valleys more easily rather than having to stay up on higher ground. These settlers would have been able to fish in the river and hunt game up in the woods above the valley. At this time the river was a marshy network of little streams occupying quite a wide river valley, nothing like the well-defined Stour of today.

The local inhabitants were part of the people known as Cantiaci or Cantii, from whom the name "Kent" was derived. The first written account of these people comes from hundreds of years later, when Julius Caesar was to write "*Of all these (British tribes), by far the most civilised are they who dwell in Kent, which is entirely a maritime region, and who differ but little from the Gauls in their customs.*"

230 BCE to 225 CE

By 100BCE Romans were exporting wine as far as Brittany, along with other Mediterranean luxuries. Cross-channel contact between the various tribes meant that Britons would have been exposed to these goods. As well as burial rites and pottery styles, Gallo-Belgic coinage was introduced and then copied by the Cantii elite.

Also by 100BCE earthwork and wood ditch-and-rampart defences had been built around Bigbury hillfort. The inhabitants were growing crops and herding cattle, and by the end of the settlement's existence the higher-status members owned slaves and chariots pulled by horses with elaborately decorated harness. A couple of miles to the northeast, another hilltop settlement (excavated in 2012) had for some time been a hive of industry with leather, textiles and metal being worked extensively. Trade links were maintained between these Cantii and other "Belgic" tribes in Gaul, probably involving boats crossing the Channel and coming up the Stour (which was navigable to within a few miles of the settlement). They practiced a druidic type of religion and maintained a tradition of bardic poetry.

After a less successful first attempt the year before, in 54BCE future Roman Emperor Julius Caesar landed on the East Kent coast with 32,000 troops, forded the river at Tonford and first met his first resistance from the locals at Bigbury, where trees had been felled to barricade its entrances. The Roman troops piled earth up against the ramparts and stormed in, only to find it empty. Caesar returned to Rome to report that Britain was worth conquering and that this could be done easily enough. But trouble in Gaul meant that the Empire suspended these plans for almost a century. In the meantime, the area was ruled by the Cantii who'd abandoned the hillfort and established a new settlement down in the valley on dry ground not far from a ford in the river.

The next 90 years saw social customs start to leak over from the Romanised mainland, with an eventual military conquest. This came in 43CE when General Aulus Plautius and at least 40,000 soldiers landed at Richborough and established control of the area with no significant resistance before heading upcountry to establish their capital at Colchester. They brought with them several local chieftains who'd been in exile, happy to collaborate with the invaders in exchange for a return to some kind of power. A Roman-style city was established beside the Stour, including some temporary military fortifications and a regular street grid lined with buildings in a hybrid Romano-British style. The city was named Durovernum Cantiacorum, meaning something like "stronghold by the alder grove, the capital of the land of the Cantii".

Although there is some evidence of resistance, the locals gradually adapted to the Roman way of life. Initially the Cantii tribal structure remained but with the chieftain now loyal to Rome. After the native queen Boudica led her failed revolt against Roman rule in East Anglia in 61CE, the Roman authorities stepped up civic development and fortification as a way of consolidating control. Regularly arranged stone and timber buildings began to replace the jumble of round Belgic huts. By 200CE the city was fully operating on a Roman model and the inhabitants of Durovernum had become citizens of the Roman Empire.

Straight roads were built (or existing ones improved) from the three nearby ports of Lympne, Richborough and Reculver, making Durovernum into a major trade hub. Roman legionnaires coming over from the mainland to serve elsewhere in Britain (or heading in the opposite direction) would often have stayed over in the city, so the provision of catering and hospitality became central to the life of the city, something which has continued for almost 2000 years. The road from Dover running SE-NW across the settlement towards Rochester and London was part of the Roman road known as Watling Street and was established as the first main thoroughfare (gravel on clay).

Both the size of the local population and its standard of living increased. It's been estimated that around 9000 people would have been living in the city by the 3rd century. It had evolved from being the main settlement of the Cantii to a prosperous community made up of administrators and merchants. Unlike Colchester this was a civilian, not a garrison, town. In the surrounding rural areas life went on much as before but self-sufficiency gradually gave way to the trading of produce through the centralised marketplaces. In town, a small class of cultured citizens lived lives of leisure enjoying underfloor heating, heated baths, mosaic floors, fashionable clothing, jewellery, imported foods and wines, boardgames, etc. Small shops and stalls sold all kinds of fruit and vegetables, oysters, leather and ceramic goods. Public and private baths were built in the city as well as a surprisingly large theatre. The Roman introduction of plumbing meant that water for bathing could be piped in from springs up in the nearby Scotland Hills.

After about a century, a Basilica and Forum were built, along with temples to Ceres, Mars, Juno, Venus, et al. Brick and tile were the primary building materials due to a lack of local stone. The theatre was expanded into a huge multi-storey structure which towered over the city and seated several thousand, one of very few Roman theatres to have been built in Britain. It's not clear what kinds of performances it was used for but it faced a large open plaza featuring a temple complex so quite possibly there was a strong religious element to them rather than just simple entertainment. This would be in keeping with what we know about the Romanised version of classical Greek drama. Most Roman towns had an amphitheatre some distance away to provide sporting-type entertainment for the rural population (and thereby provide a sense of belonging in the *civitas*) but seemingly not Durovernum.

From the various figurines and other ritual objects found in grave sites, religion appears to have involved a fusion of native and Roman gods and goddesses. Small clay figurines of the mother goddess Dea Nutrix suckling two babies would have been parts of household shrines and burials, suggesting an active cult of motherhood. Cremation eventually gave way to inhumation (outside the city). A group of four large artificial mounds on the edge of the settlement were used for burials. The construction of these would have required a huge amount of labour – whether this was done during the Roman period or earlier has not yet been clearly established.

Roman

left: some examples of late Iron Age pottery excavated within Canterbury's city walls, displaying the "Belgic" style common across the Channel at that time; *right*: the Ringlemere Cup, a finely worked ritual gold vessel, probably from 1700-1500BCE, unearthed near Sandwich.

an Iron Age gold coin dug up on the University of Kent campus during the excavations preceding the building of Turing College in 2013 – this hilltop site is now know to have been a kind of Iron Age industrial estate

left: a "hooded dwarf" figurine dating from the early days of Durovernum Cantiacorum and dug up from what appears to have been a temple under St. Gabriel's Chapel in the Cathedral – it was originally coated in mica gilt and is believed to be linked to an Aesculapian cult which existed in Northern Europe in the early centuries CE; *right*: a figurine of Dea Nutrix, the Romano-British mother goddess, several of which have been found in burial sites around the city

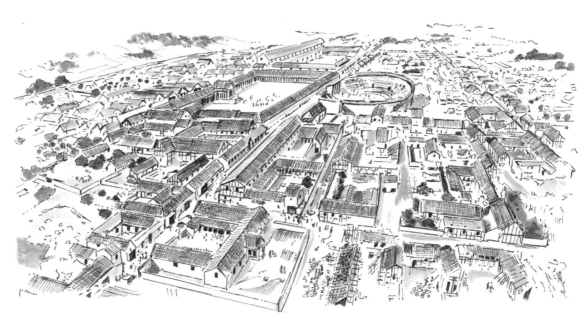

a reconstruction of Durovernum around 150CE with the original theatre in place, facing a plaza and temple complex

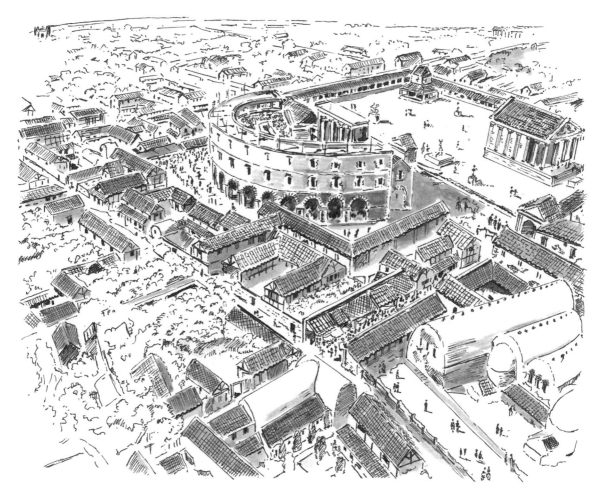

a reconstruction of Durovernum around 300CE with the new amphitheatre having replaced the original theatre

225 to 589 CE

In the early 3rd century the main thoroughfare had changed from Watling Street to a route entering the city from the southwest at the same point (the future "Ridingate") but at a slightly different angle, exiting at what was later to become the Westgate rather than the future London Gate.

A gradual economic decline occurred in the 3rd century. As across most of the Roman Empire, urban life became more expensive and this caused the city's population to decline. Luxurious houses fell into disrepair and many buildings were left unoccupied. Roman ruins in the northwestern part of the city have been found waterlogged, so there appears to have been major subsidence there at some point during the Roman occupation. The branch of the river which now runs within the city walls was possibly cut as a drainage channel, draining the marshy network of streams which made up the Great Stour at that time.

By the end of the 3rd century, Saxon bands were attacking nearby coastal villages. In response, several Roman forts were built along the coast between 290 and 300. This involved considerable cost, further weakening the local economy. Tall stone walls several feet thick were built around the city between 270 and 290, backed by earthen ramparts and with eight gates. These walls enclosed a much bigger area than that covered by the existing dwellings, thus providing room for expansion and space for the nearby rural populace to shelter in case of attack. But it's been suggested that, as with Rome, the wall may have been at least as much of a status symbol as a fortification. In any case, an industrial suburb where metalworkers, potters and tile makers had been thriving on the other side of the river began withering away, having been cut off from the rest of the settlement. By the early 4th century, Durovernum had effectively become the trading centre for a vast armed camp. Defending the coast from Saxon attack became the city's main concern, so traffic passing through began to be more of a military than a commercial nature.

In 313 the Roman Emperor Constantine issued an "Edict of Toleration" after which the previously outlawed, relatively new religious cult of Christ was now officially tolerated (Constantine was to convert to Christianity shortly thereafter, making it the official Imperial religion). Christianity had arrived in Durovernum, brought by Britons returning from military service in other parts of the Empire, and seemingly had been operating underground for some time, as the city's first Christian church sprung up shortly after the Edict. With Christianity, the burial of ashes along the sides of roads out of the city was gradually replaced by inhumation in consecrated ground, eventually within the walls. Burial sites inside the city walls were appearing from the end of the 4th century. Possibly in response to Christianity becoming the state religion, a pagan counter-current sprung up briefly in the city, with an octagonal stone temple being built in 340 and lasting until about 360. A temple abandoned sometime before 300 later became the site of St. Gabriel's Chapel in the Cathedral crypt where archaeological digs were to turn up a mosaic floor, a bull's horn, a bronze representation of the Graeco-Roman wine god Silenus and a mysterious hooded dwarf figure.

In 407 Emperor Honorius ordered forces to be withdrawn from the coastal forts and to return to defend the core of the Empire from Goths, Vandals, etc. By 410 the Roman army had left Britain, leaving the people of Durovernum to defend themselves. According to the *Anglo-Saxon Chronicle*, Durovernum was abandoned when the Saxon mercenaries Hengist and Horsa turned on their British warlord employer Vortigern and attacked the city. Historians generally don't accept this, as there's no archaeological evidence for widespread destruction at that time. But there is strong evidence from the soil layer that the settlement was almost entirely abandoned for most of a century. This could have been entirely due to economic collapse, with soldiers' salaries no longer in local circulation. Coins ceased to be minted, the upper middle classes left the city to become rural villa dwellers and the place became a ghost town.

Whether invaders or peaceful immigrants, Saxons weren't interested in urban living, so the walled city fell into desolation with only a tiny population remaining. The walls began collapsing, due to neglect rather than attack, and population centres began to emerge elsewhere in the area. By the early 500s, though, another wave of Northern European immigrants began to resettle the inside of the city walls with their small wooden houses. It appears that Saxons, Jutes, Angles and Frisians settled, coexisted and intermarried with each other and the existing population. A new, more organic, street plan began to emerge, the Roman one having been lost below a layer of soil. By 560 the new settlement "Cantwaraburgh" (also rendered "Cantwaraburh", "Cantwaraburg" or "Cantwarabyrig") was the capital of Kent, soon to become the most powerful of the Saxon petty kingdoms. By the 6th century, via Saxon settlement in the north, new ideas about social organisation were influencing southern Britain and small kingdoms were emerging. Due to its control of most trade between Britain and mainland Europe, Kent was to exercise power over kingdoms as far north as the Humber. The "burgh"/"burh"/"burg"/"byrig" in the new name indicates a fortified settlement and modern street names (Burgate, The Borough) suggest that this would have enclosed an area within the city walls equivalent to what eventually became the Cathedral Precincts. The entire circular walled area was presumably too big to plausibly defend, hence some kind of earthwork/fence arrangement enclosed a more manageable eye-shaped portion of it.

Around 589 Ethelbert became King of Kent, beginning an apparently peaceful reign during which various social institutions emerged. Around 580 he'd married the Frankish princess Bertha, on her family's condition that she could continue to practice her Christian faith. Bringing her chaplain Bishop Liudhard with her, she was given an old Roman building (possibly a temple) outside the city walls to use as a church. This became St. Martin's, used for Christian worship to this day. Questionable tradition (via the Venerable Bede) says that this occupied the same building as the church that had sprung up after Constantine's Edict tolerating Christianity. If so it would have been paganised in the interim as Anglo-Saxon settlement had effectively stamped out late Roman Christianity. Ethelbert and Bertha's Cantwaraburgh palace stood close to the future Cathedral site, so she would have taken a daily walk to St. Martin's through the Roman gate that consequently became known as the Queningate (the Queen's gate). On her way she would have passed a pagan temple which may well have been used by her husband to worship his Saxon deities.

589 to 878 CE

In 595 Pope Gregory decided to send an envoy from Rome to convert the pagan people of Britain, seen as a "lost province" of the Church which had established a foothold there in the 4th century before its Anglo-Saxon suppression. He chose the monk Augustine, prior of one of his monasteries, sending him with instructions to establish two provinces based at the former Roman administrative centres of London and York. After an initial turning back, daunted by the task of converting the Britons, Augustine and his contingent of Frankish clergy persevered, arriving at Ebbsfleet (near Pegwell Bay) around Easter 597. The pagan King Ethelbert was open to the idea of Christianity, having been married to Bertha for many years, and may have been wanting to convert. In fact he may even have quietly invited the Pope to send a delegation, wanting to convert in style. Having travelled to the coast to meet Augustine and his company, he was sufficiently comfortable with the missionaries to invite them to establish a base in Cantwaraburgh, so a procession made its way into the city, led by Augustine holding up a silver cross and, according to tradition, chanting "Let Thy anger and wrath be turned away from this city".

The Papal delegation began operating from Bertha's church of St. Martin and Ethelbert soon gave Augustine a site near the royal palace to build his cathedral (possibly a former Roman temple to convert). He had himself baptised by about 601 and moved his base north to Reculver, so within a matter of decades Cantwaraburgh had switched from being a political capital to a spiritual capital. Pope Gregory's plan to establish the archbishopric at London never came to pass, it having stayed at Canterbury ever since Augustine made himself the first archbishop (originally "Metropolitan Bishop") in 597. A second team of missionaries arrived from Rome in 601 with vestments and illuminated books, one of which is still used in archbishops' enthronement ceremonies.

Having established Christchurch Cathedral with an associated priory, in 598 Augustine founded a second monastic institution just outside the city walls, an abbey dedicated to St. Peter and St. Paul. Its primary purpose was to provide a consecrated burial ground for archbishops, as burial within the city walls was again seen as unacceptable (so burial within the Cathedral was not an option). A stone church there was completed and consecrated by 613. Building materials were recycled from old Roman buildings. The pagan temple on Bertha's route from her palace to St. Martin's was enclosed within the Abbey grounds and converted to a chapel dedicated to the Roman child martyr St. Pancras.

Around 616 Ethelbert died and was succeeded as king of Kent by the pagan Eadbald. A brief anti-Christian backlash followed and the second Archbishop of Canterbury, Laurence, considered fleeing. He stood his ground, though, and Eadbald converted soon after. After Ethelbert's reign the political importance of Kent began to wane. Various kingdoms gained supremacy over it in the centuries that followed but all were Christian so the institution of the archbishopric remained intact throughout. The early archbishops were all Roman monks but fulfilled a kind of "tribal wizard" role for their Saxon communities, the first ten of them becoming saints after death and their tombs attracting cults (the Saxon cult of saints acted as bridge into Christianity from their earlier faith which involved a multiplicity of gods and goddesses associated with various aspects of reality). The canonised archbishops' shrines were seen as places of healing where the bones of the saint somehow mediated between Heaven and Earth. Monks of the Abbey and the Cathedral Priory were seen as custodians of the saints' earthly remains and the living archbishop personified his sainted predecessors. The first Archbishop of Canterbury to be of English birth was Deusdedit, enthroned in 655.

By the end of the 7th century there was a flowering of learning at the Abbey, the school founded there by Archbishop Theodore of Tarsus and Abbot Adrian being considered at the time to be the greatest in Europe. Canterbury began to send missionaries out to mainland Europe and beyond. Historians consider the school to be one of the pinnacles of intellectual culture in the Early Middle Ages. On the basis of the school's merit Pope Agatho (678-681) gave the Abbey monks exemption from the authority of the Cathedral and freedom to elect their own abbot. This eventually led to significant tensions developing between the Abbey and the Cathedral.

Much of the inside of the city walls was still unoccupied or used for farming but huts were being built in the city centre, clustered around the ruins of the Roman theatre. Unlike the round Belgic huts of the Cantii, Saxon huts were rectangular, sometimes making use of remaining Roman walls or street alignments. A new commercial thoroughfare emerged, roughly at a right angle to the Roman one, running from the Worthgate to the Cathedral. Weaving, leatherwork and pottery making were all going on, gold coins were being minted and goldsmithing was producing some remarkably intricate jewellery. References to active ports at Fordwich and Sandwich start to appear in documents around 700.

By 750 there were nearly twenty churches in the city. Around this time, Archbishop Cuthbert built yet another one almost touching the east end of the Cathedral, dedicated to St. John the Baptist. He was buried in this, breaking the tradition of Abbey burials, and it subsequently became the new burial place for archbishops. Before this, the Abbey was seen as taking precedence over the Cathedral, as it housed the remains of six sainted archbishops. A power switch occurred at this point, the Cathedral starting to become the dominant institutution. In the early 9th century both the Abbey and the Priory began to acquire estates around Southeast Britain, leading to financial independence but also to royal resentment.

Coins had been reintroduced by King Ethelbert and by the 9th century seven mints were operating in the city. 835-855 saw a wave of Viking attacks on Kent with attacks on Canterbury occurring in 842 and 851 ("great slaughter" was recorded in the latter instance although few details are known). By 878 Vikings were controlling half of what was to become England with York as their capital. Due to the continual threat of Viking attack, the Abbey school had gone into serious decline, learning and the production of manuscripts having reached a low point. But the cultural life of the city continued, as illustrated by an exquisite bronze brooch made around this time (the "Canterbury Cross", found beneath St. George's Street in 1867).

a reconstruction of the Roman Ridingate, where Watling Street (the road from Dover to London) entered the city walls

a reconstruction of the original St. Martin's Church where Queen Bertha worshipped in the late 6th century and which St. Augustine used as his missionary base (adapted from an earlier Roman building, some bricks of which survive in the current building, the oldest continually used place of Christian worship in the English-speaking world)

a cache of silver dug up near the Westgate, dating from about 410CE and including an implement with a Christian "chi-rho" symbol (believed to be for dividing communion bread), evidence that Christianity was being practiced in the city at the time

an artist's impression of the amphitheatre in a state of ruin some decades after the Roman Empire abandoned the city

an impression of the resettlement of the inside of the city walls in the early 6th century by Northern European migrants (Jutes, Saxons, Angles and/or Frisians), as viewed from near the Ridingate

left: the 9th century "Canterbury Cross" found beneath St. George's Street in 1867, eventually to become a symbol of the worldwide Anglican communion; *right*: a Saxon-era gold pendant dug up in a cremation cemetery on the London Road in 1982

878 to 7th January 1109

By the late 9th century, monks at the Abbey, once a great centre of learning, were barely literate. But Canterbury remained an important centre of power – by 930 it had seven mints compared to London's eight: four royal, two associated with the Cathedral and one with the Abbey. Having come from Glastonbury Abbey, the celebrated Archbishop Dunstan (who held the office from 959 to 988) found the Canterbury monks to have become corrupt and lax, so brought about extensive monastic reforms, introducing the Benedictine rule. Dunstan became exceptionally popular among Archbishops of Canterbury, being made a saint and his shrine in the Cathedral becoming the leading pilgrimage destination in England for a couple of centuries.

Between 991 and 1012 Canterbury had yearly trouble with Vikings who were seeking to plunder Christian institutions to fund the wars being fought against them by (nominally Christian) Germanic tribes. In 994, Archbishop Sigeric "The Serious" saved the Cathedral by paying the Danish raiders protection money. In 1002, King Aethelred took the drastic step of ordering the murder of all Danes in his kingdom, leading to major recrimination from across the North Sea. In 1009, a raiding party agreed not to destroy the city for a fee of £3000. Two years later the Vikings, led by Thorkell the Tall, returned to besiege Canterbury for three weeks. This culminated in the burning of the (wooden) Cathedral, the murder of possibly thousands and Archbishop Alphege being taken hostage for an exorbitant ransom. The Abbey was spared, having provided food and shelter to the aggressors (arguably with little choice). Some months later in Greenwich, Alphege was murdered at a drunken feast after refusing to allow a ransom to be levied on his people, thereby becoming a saint and a national hero.

By 1016 England was under the control of the recently converted Christian King Cnut, a Dane who was crowned by Archbishop Lyfing and who paid for the rebuilding of churches and cathedrals destroyed by his people. In 1023 he arranged to have Alphege's body returned for burial at the Cathedral (plus gifts and land grants including the nearby port of Sandwich). St. Alphege's shrine became another pilgrimage destination, good for the Cathedral coffers and the city's economy. By 1042 the Cathedral had been rebuilt, with a new oratory of St. Mary. Because the two most important religious institutions in the country were located here, Canterbury had also become commercially prominent. At this time it was divided into three zones, under the jurisdiction of the king (via an official called his "portreeve"), the archbishop and the abbot. Each had its own mint and its own markets, sometimes leading to disputes over rights to extract tolls from outside merchants. Most of the original Roman gates were still being used, the Burgate probably being the main entrance into the city during this time. Markets were mostly held outside the city walls with an Abbey market in Longport and a Cathedral market possibly outside the Westgate.

Danish rule ended with the death of Cnut's son Harthacanute and the return of the Saxon line with Edward the Confessor. When Edward died heirless in 1066, the national Saxon witan (assembly) elected Harold, son of Earl Godwin, to the throne. Duke William of Normandy, being Edward's second cousin and claiming to have been promised the kingdom, invaded later that year. He had the blessing of Pope Alexander II after a dispute involving Harold having appointed Stigand as Archbishop of Canterbury in 1052 without papal approval while he and his father Godwin had been administering the country on Edward's behalf.

After defeating King Harold's forces at the Battle of Hastings, William's men went on to attack Dover before heading up to Canterbury. With the Viking siege of 1011 still in living memory, the city willingly surrendered. Canterbury was of strategic importance as a place to ford the River Stour, so William ordered a castle built, a small round wooden fortification atop the largest of the group of four (possibly pre-Roman) mounds which was now just inside the city wall. That mound then became known as the "Don Jon", later "Dane John", a corruption of the old French word *donjon*, meaning a castle keep. William continued on up to Rochester and London where two more castles were built (controlling crossing points of the Rivers Medway and Thames).

The religious life of Canterbury continued without too much disruption, although William's men had seized many estates belonging to the Cathedral and Abbey on their way through Kent, undermining their finances. In 1067 a fire completely destroyed the Cathedral (recently rebuilt, again in wood) bringing to an end its Saxon phase. It was probably the largest church in the country at that point and seen as an imitation of St. Peter's in Rome with Canterbury as a kind of mini-Rome in England, the monks following the Benedictine rule and Roman liturgy. In 1070 Stigand, the last of the Saxon archbishops, was deposed and replaced by William's choice, Lanfranc. The charismatic/mystical/monastic Saxon model of the Church in England had focussed on the cult of saints, including many early archbishops and, prominently, St. Mary. A shift then began to occur towards the more territorial and bureaucratic mode that was prevalent on the European mainland. But the spirit of the Anglo-Saxon church continued on in the form of monastic life.

Lanfranc was enthroned in a temporary wooden building (the Cathedral still being a charred ruin) and immediately set to work repossessing Christchurch estates seized after the Conquest. He also started to have Caen stone (a type of Jurassic limestone) shipped in from Normandy to build an entirely new Cathedral, this being completed by 1077. He increased the number of Priory monks from 50 to 150, instituting reforms after having found them to have again become decadent, hunting, gambling and drinking heavily. A couple of hospitals – in the original sense of "places of hospitality" – were built to care for the elderly and lepers, respectively, just outside Northgate (St. John's) and in the village of Harbledown just below Bigbury hillfort (St. Nicholas's), as well as a new Priory of St. Gregory just across from St. John's Hospital and an Archbishop's Palace within the Cathedral Precincts.

Beyond the life of the religious houses not much changed in the life of the city. Freemen, a class of adult male residents recognised as having a certain set of civic rights, swore allegiance to the new king and paid higher taxes. By 1085, a new stone castle was being built near the Worthgate to replace the small original castle atop the mound. The position of this suggests that it wasn't built to defend the city from outside invasion but rather to defend the Norman king and his men from the the local population when he was staying in the area.

Early Medieval

17th January 1109 to 30th November 1292

The king's sector of Canterbury was being run by a royally appointed governor. By the 13th century this had given way to a system involving two bailiffs, still appointed by the king but now chosen from the freemen of the city. The city had been divided into six wards, each named after the nearest gate and each with an alderman and a constable (the city continued to have aldermen until 1974). Later, Henry III granted a charter to the city allowing the people to elect their own bailiffs and handing over responsibility for raising taxes, as well as control of the walls, gates, buildings and rents, in exchange for a yearly lump sum of £60.

Most residents were farming small strips of land outside the city walls or grazing sheep in common fields. The downlands south of the city provided excellent sheep pasture, boosting local wealth. The feudal system that was now well established had created sharp economic divisions between different classes of inhabitants but there was also a distinct power gap between the religious and civil communities. The Cathedral was still minting coins and lesser clergy were involved in commerce, with a Guild of Churchmen formed to regulate their activities. Christchurch Priory had its own area of jurisdiction, conducted its own trials and had its own jails, pillories and gallows.

In 1161 fire destroyed a large part of the city. By this time extensive suburbs had spread beyond the Northgate, Westgate and Worthgate, shops lined Burgate Street and there were at least thirty stone houses within the walls. The walls were rebuilt and/or built up during this era. The first records of Jews settling in the city appear in the late 12th century. As in the other trade hubs of the kingdom, they provided moneylending services that were otherwise banned by usury laws (laws whose origins lay in the Old Testament). Although their clients included the Cathedral and Abbey they were still sometimes fined for usury.

Arguably the pivotal event in the entire history of Canterbury occurred on 29th December 1170. Thomas Becket was an archbishop who'd been appointed by Henry II in 1162. To Henry's dismay, his former clerk and close friend took the role very seriously and began defending the Church's privileges against the Crown, exactly the opposite of what the King had intended. Consequently, Becket spent most of his eight years as archbishop in French exile. After extensive excommunications, threats, recriminations and negotiations, Thomas returned to a hero's welcome in Canterbury (during his brief initial time in office he'd adopted a life of poverty and spent much of his time caring for the poor and sick). Stories from the time describe hundreds of people wading out into the sea off Sandwich to meet his ship and the road to Canterbury being lined with people from around Kent who'd come to receive his blessing, some throwing themselves at his feet.

After twenty-seven days back at his Cathedral, four of the Kings knight's arrived, armed, to confront him. It's unclear whether they were ordered to do so by Henry or were just trying to impress the King to gain favour. But the events that followed are the best documented in medieval history, culminating in Thomas's murder inside the Cathedral. His horrified monks quickly and carefully mopped up his spilled blood and brains, and reports of miraculous healings occurring through contact with the blood-soaked rags soon began to circulate in Canterbury and beyond. Before long, the murdered archbishop had become a saint and martyr, the resting place of his bodily remains attracting a stream of pilgrims.

In 1174 King Henry himself made pilgrimage, attempting to make amends for his role in the affair. Having stopped to make offerings at St. Nicholas's leprosy hospital in Harbledown and continuing on to St. Dunstan's Church, he put on a hairshirt under a pilgrim's cloak and walked barefoot into the city. After praying at Thomas's shrine he voluntarily had himself whipped by each of the monks in turn, spent the night bruised and bleeding in the Cathedral crypt and returned to London the next day, seriously ill.

Later that year another fire destroyed large parts of the Cathedral, but Becket's tomb in the Crypt survived. The rebuilding programme, under the direction of William of Sens (and later William the Englishman), led to perhaps the finest work visible in today's Cathedral: the Trinity Chapel and Corona Tower. A few years after the fire, Gerald of Wales visited the Priory and was shocked by the decadence of the monks – their luxurious meals, lack of modesty and frivolous behaviour. In July 1220 Becket's remains were ritually "translated" to a newly constructed shrine in the recently completed Trinity Chapel. The cult of Thomas had grown to such an extent that this was a major event attended by important nobility and clergy from all over Britain and Europe. The city and surrounding villages were overflowing with visitors and it took four archbishops to pay off the loans required to stage this lavish exercise of devotion to St. Thomas of Canterbury.

Conflicts between kings, archbishops and monks dominated this period of the city's history. In the 1180s Archbishop Baldwin attempted to establish a "College of Cardinals" rather like the one in Rome which elects popes, just north of the city in Hackington. The Priory monks resisted this fiercely, fearing it would undermine their rights and privileges. The conflict culminated in the monks being besieged in the Priory for some months, with locals having to smuggle food in to them. Baldwin's scheme ultimately failed but he did begin a building project at Lambeth, on the south bank of the Thames, eventually to become the primary residence of future Archbishops of Canterbury. Work on this was continued by Archbishop Hubert Walter. When Walter died in 1205, King John and the Priory monks clashed over who should succeed him. The Pope became involved in the dispute and ended up choosing his own candidate, the English cardinal Stephen Langton. The enraged King reacted by declaring the monks to be traitors and driving them into exile in Flanders. The Pope reacted by placing the whole of England under an interdict for six years, forbidding church services, baptisms, weddings and burials in consecrated ground. Around 1240 Archbishop Edmund fell into conflict with the monks over their rights and customs, leading him to excommunicate them en masse. When they chose simply to ignore him, he spent the rest of his life at Pontigny in France. After the establishment of Lambeth Palace, just over the river from the Palace of Westminster and Westminster Abbey (where kings were crowned, married and buried), archbishops began to spend less and less time in Canterbury, having been drawn into the orbit of London, the national centre of politics and power.

High Medieval

Conflict between the citizens and the Abbey appears to have come to a head in 1257. A local woman was taken into custody and jailed by the Abbey, being accused of a felony committed in the adjacent Longport area. The jurisdiction of this area had been vigorously disputed for some time it seems, as the bailiffs raised a large crowd of commoners by blowing the ritual Burghmote Horn and then led an attack on the Abbot's Mill. Millstones were overturned, equipment damaged and the miller and his servants injured. The next year, the disputed jurisdiction was addressed in a document jointly signed by city and Abbey authorities to clarify such matters.

As well as the Christchurch monks, King John came into serious conflict with his barons, noblemen who owned estates in both England and France, owing no particular allegiance to the English Crown. Having successfully pressured him into signing the Magna Carta (a kind of treaty between the barons and the Crown) in 1215, the barons reacted militarily to John's refusal to abide by its terms, leading to a couple of years of civil war. During this time King Louis VIII invaded the south of the country and had himself proclaimed King of England before being ejected the next year. So, for part of 1216, Canterbury's royal castle was under French occupation. Life went on relatively unchanged for the locals and religious institutions, although the minting of coins was suspended and Archdeacon Simon Langton (Archbishop Stephen's brother) got directly involved in Louis's regime, becoming his chancellor during the brief occupation. This displeased the Pope enough to get Langton temporarily excommunicated. But because he also indirectly helped to fund a small boys' school in the city (which has continued to exist in some form since its foundation), rather than being remembered as a French collaborator, Simon Langton's name is today associated with a prestigious grammar school in Canterbury which was to produce some fantastic musicians some 750 years later.

King John's son Henry III had more trouble with the barons in the 1260s. A second civil war involved a baronial army led by the French-born Simon de Montfort sweeping across Kent (where they found much support) and capturing Canterbury in 1264. That August saw Henry in the city, being forced by the barons to sign the "Peace of Canterbury", an arrangement which was highly unfavourable to him. By the next year, though, Montfort was killed and full royal rights were restored.

When Gilbert de Clare led the Baronial army into Canterbury, one of the first things to happen was an attack on the Jewish quarter. The city's Jewish population had been steadily growing since the end of King John's reign in 1216, concentrated in the city centre on the opposite side of the main thoroughfare from the Cathedral. Their role in moneylending and the wealth that this brought them caused tension and in 1222 Archbishop Stephen Langton had attempted to pass ordinances prohibiting associating with Jews, selling them food and the building of more synagogues. This was overruled by one of young King Henry's protectors, the Keeper of Canterbury Castle, as the Crown relied on their loans and stood to benefit from the special taxes it levied on Jews in the Kingdom. In 1261 Canterbury had seen an attack on Jews' houses by a band of angry men. Windows and doors were broken, houses were set on fire and there were violent assaults, but no deaths. The King then ordered the Sheriff of Kent to investigate and punish the perpetrators. The baronial attack on Jewish houses in 1264 was motivated by a desire to capture the *Archa*, a chest full of handwritten duplicates of bonds, evidence of outstanding debts owed by Christians to Jews. This was eventually found in the house of Bailiff Simon Pabley, which had been forcibly entered by baronial soldiers. It was taken to Dover, never to be seen again.

The Jewish quarter was again attacked in 1276, a few years into Edward I's reign. A couple of years later, Edward charged every Jew in the Kingdom with coin-clipping and Canterbury's Jews were imprisoned in the castle (Hebrew graffiti was reportedly discovered there by an antiquarian in 1674). In 1290 all Jews were expelled from the Kingdom, so the area around Jewry Lane from St. Mary Bredman to All Saints Church became deserted. They sought refuge across the Channel while the King gave away their houses to his friends.

In the early 13th century new religious orders were emerging in mainland Europe and these inevitably made their way to the seat of English Christianity. Throughout the Kingdom large abbeys and priories had become very wealthy indeed and there would have been resentment of St. Augustine's Abbey and Christchurch Priory among local people who could see how far they'd drifted from their original mission. The new orders of "friars" seemed much more appealing in comparison. They preached directly to the people rather than mystifying them with Latin chanted from a distance, stayed closer to their vows and encouraged a more active style of belief. The Dominicans or "Blackfriars" were the first to arrive, in 1220. They impressed Archbishop Langton and some years later were given a patch of land spanning the river (including an island that's no longer an island). Caring for the poor and sick, they were well liked locally. Arriving a few years after the Blackfriars (but the first to establish themselves in Canterbury) were the Franciscans or "Greyfriars". They were also given a damp parcel of land beside the Stour, further upstream and including the island of Binnewith. In the 1250s an extremist splinter group of Blackfriars known as the "Friars of the Sack" established themselves in a house beside the imposing black flint Blackfriars Gate in St. Peter's Street.

Around 1220 Alexander Frese converted his stone house in Stour Street into an almshouse for poor priests, dedicating it (like so many of the churches in the city) to the Virgin Mary. Alexander had acquired this house from his father, Lambin the Minter, who had been forced to relocate his premises there from a location near the Cathedral after the fire of 1174, an event he may have been responsible for. The Poor Priests' Hospital was formally founded by Archdeacon Simon Langton in 1240. The previous century had seen the establishment of hospitals of St. Lawrence (Dover Road, for lepers and their families) and St. Thomas (beside the Eastbridge on the High Street, for seriously ill pilgrims).

By 1270, the fifty year jubilee of St. Thomas's "translation", pilgrimage to Canterbury was in full swing. Mercery Lane was lined with merchants selling cheap souvenirs: lead badges featuring Becket's head and ampullas of supposed holy water. In 1272, Canterbury experienced terrible thunder and lightning storms. The city was badly flooded, trees were uprooted and herds and flocks were driven out of nearby fields leading to food shortages, disease and widespread death.

High Medieval

left: the original "motte-and-bailey" castle which William, Duke of Normandy, ordered built (atop the mound which would come to be known as the "Dane John") when he rode through Canterbury during his 1066 conquest of England; *right*: the much larger stone castle which soon replaced it nearby (not so much to defend the city as to defend the visiting king from the citizenry), the ruined keep of which still stands

the Burghmote Horn, the oldest piece of civic regalia still used by Canterbury City Council in the 21st century, dating back to at least the 12th century (if the unbroken tradition is kept, it will be blown to announce the enthronement of the next monarch)

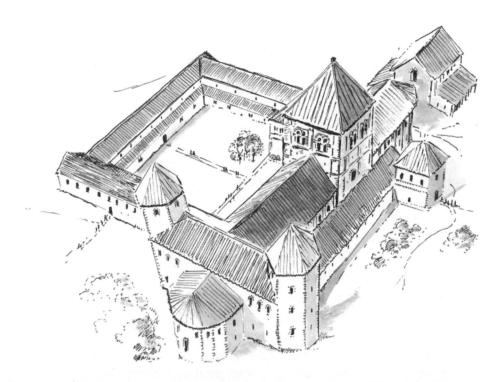

the final (wooden) Saxon Christchurch Cathedral/Priory as it would have looked in 1067 just before it burned down

left: the early 14th century Fyndon Gate to the Abbey grounds, named after the Abbot of St. Augustine's at the time; *right*: the Poor Priests' Hospital in Stour Street, built 1175-1180 and originally the home of the minter Lambin Frese (this is an impression of the version rebuilt in 1373) – it's still standing, housing the Canterbury Heritage Museum, after having been seized by the Crown in the 1530s, given back by Elizabeth I in 1574, then been variously a "house of correction", a workhouse, a school and a police station

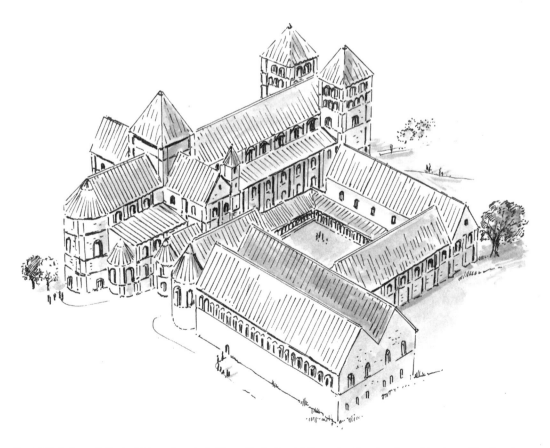

the original stone Cathedral/Priory, largely built mostly in the 1070s under the direction of Lanfranc (the first Norman Archbishop of Canterbury), as it would have looked around the time of his death in 1089

30th November 1292 to 14th June 1439

Between 1220 and 1348 the city's population increased from 6000 to 10,000, only to be then severely reduced by the Black Death. The population of England was almost halved, with Canterbury and Kent particularly badly hit. Two archbishops were victims of the plague, although the Christchurch monks weren't as badly affected as the rest of the city thanks to an excellent plumbing system introduced by Prior Wibert nearly 200 years earlier. The reduction in population led to a decline in agriculture, affecting wages and prices in a way which both hurt the finances of the Cathedral and Abbey (who owned large estates around the southeast of the country) and was eventually to lead to the Peasants' Revolt, where anger was directed at both the Crown and the Church, who were among the worst landlords. It also led to a rearrangement of the parish structure in the city, with St. John's Parish being absorbed into that of St. Mary de Castro (a church originally built to serve the castle) and St. Edmund Ridingate merging with another parish. Within a few decades, St. Mary Queningate was gone too, possibly absorbed into the Cathedral Precincts when the walls were rebuilt. But veneration of the Virgin was still very much active in the city, with a new shrine of "Our Lady Undercroft" being constructed in the Cathedral Crypt.

Alongside catering for pilgrims, the weaving and dying of wool became Canterbury's major industry. Initially, raw wool was shipped out from Fordwich to Calais but soon Flemish weavers were encouraged to settle and more valuable manufactured woollen items were being exported. The combination of the wool trade and the flood of pilgrims to Becket's shrine led to an economic boom between 1350 and 1420. The adult male population was divided between freemen and intrantes ("enterers"), the latter having to pay an annual fee to live and/or trade in the city. To be a freeman required either being born of a freeman, married to one's daughter or having paid for the privilege with the consent of the Burghmoot, a predecessor of the city council. Certain wards lay outside the city's jurisdiction when it came to the matter of freemen and intrantes. Westgate ward was under Christchurch jurisdiction, while St. Paul's was under the Abbey's. Intrantes would sometimes move between wards to avoid paying fees.

In 1380 the Roman Westgate was replaced by the current one, involving a pair of cylindrical ragstone towers and funded by Archbishop Simon Sudbury. By 1400 the city walls had again been rebuilt (partly at the expense of Archbishops Chillenden and Selling), this time with multiple towers, some with gun ports. Henry IV granted more power to the city's bailiffs, including the right to buy property. In 1408 they bought and significantly altered the Red Lion Inn which stood next to the council chamber known as the Moot Hall. The Red Lion was to become the city's "talking shop" for many centuries. By 1427 the Moot Hall was being referred to as the Guild Hall (due to its use by the numerous guilds of craftsmen in the city) and the Westgate had become the city jail, the castle being the county jail.

Skeletal analysis indicates that the life of Canterbury's commoners during this time regularly involved rickets, scurvy, arthritis, anaemia and bad teeth. Not surprisingly, infant mortality rates were very high. The state of public health was well represented by the notorious "black ditch", an open sewer running through the Don Jon Field (later to become the Dane John Gardens). Things were made yet worse for the people when a powerful earthquake shook the city in May 1382. This was strong enough to bring down parts of the Cathedral's bell tower and cloister walls as well as damaging many other buildings in the area.

Numbers of pilgrims to Canterbury had been steadily growing, especially since the translation of Becket's remains to their new shrine in the summer of 1220. Every fifty years thereafter a special jubilee was declared, including papal indulgences. Although pilgrimages were made all year round, the July festival of the translation was the most popular time for them. The 1420 jubilee was probably the peak year for pilgrim numbers, some records claiming that 100,000 visitors were in and around the city for it. This was almost certainly an exaggeration as it's hard to imagine how the city's few thousand residents could have catered for such an influx. In any case, it's certain that the place was packed with pilgrims, the inns overflowing and poorer visitors camping outside the walls or just sleeping rough. The Priory had a number of pilgrims' inns built, most famously the Sun Inn (still standing in Sun Street) and the Chequer of Hope Inn (still partly standing on the corner of Mercery Lane and the High Street).

Although not published until 1478, Geoffrey Chaucer's *The Canterbury Tales* was written in the late 1380s or early 1390s, describing a group of pilgrims making the journey down from London but ending in Harbledown just before they reach the city. There's no material evidence that Chaucer ever set foot in Canterbury but as a bureaucrat and diplomat moving between London and mainland Europe it's very likely that he did. The *Tale of Beryn*, an anonymous "sequel" to his *Tales*, appeared not too long after, perhaps the earliest example of "fan fiction". It described Chaucer's pilgrims staying at the Chequer of Hope, visiting Becket's shrine, stealing souvenirs and generally behaving badly.

As the popularity of the pilgrimage to Canterbury increased, its religious significance declined, many pilgrims either seeking miraculous cures for their ailments or just a pleasant summer holiday. The word *canter* appears to have originated during this time, describing the gait of a horse between a trot and a gallop. People came to speak of "the Canterbury pace" – not too slow, as you wanted to reach your destination, but not too fast, as you wanted to enjoy the journey – and this eventually got shortened into the two-syllable word. The majority of traffic was on the Pilgrims' Way from London via Rochester, with pilgrims entering the Westgate, proceeding down the high street, turning down Mercery Lane (then covered) and through Christchurch Gate to be met by monks and shepherded to the shrine of St. Thomas. There they would kiss the reliquary chest before venturing out into the city to buy souvenirs, drink ale and otherwise enjoy themselves. Less widely used pilgrims' routes ran from Dover (now the area's main port) and Winchester. As well as the throngs of pilgrims, Canterbury was accommodating clergy, diplomats, lawyers and merchants on their way from London to the European mainland (or vice versa). It had now truly become something like a little Rome in England, its significance now extending well beyond national borders.

Both the religious and civil sectors of Canterbury benefited greatly from the pilgrimages so an uneasy truce overrode the simmering rivalries which had built up over the centuries. But there were still eruptions of trouble throughout this time. During the first half of 1297, the Canterbury clergy were outlawed by King Edward for refusing to contribute sufficiently towards his military campaigns. Edward had asked for one-fifth of their income but Archbishop Winchelsey had refused to cooperate without first getting the Pope's agreement. The Sheriff of Kent's men entered the Priory and shut down the domestic quarters – being outlawed meant forfeiting all possessions. The Archbishop was forced to live off charity and even had his horses confiscated when he travelled to meet the King. The situation was resolved when his monks disobeyed him and paid the Crown the amount demanded. A similar situation arose in 1327 when there were disturbances caused by the Christchurch monks' refusal to contribute towards Edward III's campaign against the Scottish. Representing the citizenry, Bailiff William of Chilham led a demonstration in the Blackfriars' churchyard (a place used for public meetings and preaching), threatening violence against Christchurch property and a boycott of the monks. There was even a veiled threat to loot Becket's shrine: a promise was made that every commoner would wear a ring of gold belonging to St. Thomas. Violence was avoided when the monks agreed to a compromise.

The friars and local parish priests often sided with the locals in disputes with the Priory and Abbey, a rebellion not against the faith but rather against their bad practice as landlords in the city. In 1314 the breakaway Blackfriars known as the Friars of the Sack were banned and dispersed after about seventy years in the city, their house in St. Peter's Street being offered to a newly arrived group of Augustinians or "Whitefriars" a few years later. The Whitefriars preferred a location for their chapel in St. George's Parish where they paid no rent to either the parish or the Priory, although that situation had been ironed out by 1325. By the early 14th century the Northolme area just north of the Abbey had come to be seen as a den of criminality and fornication, so Abbot Ralph de Burne closed the common road, cleared the ground of thorns and brambles, and planted a vineyard, thereby "ending the district's evil reputation".

More monastic corruption of a kind which would certainly have alienated the citizens of Canterbury was reported in 1338 by the new Archbishop, John Stratford, who expressed a concern for lowering moral values among the Priory monks. He claimed that they were not wearing clerical clothing but rather short, tight tunics and "*boots of red and green, peaked and cut in many ways*". They "*distinguish[ed] themselves by hair spreading to the shoulders in an effeminate manner*", "*their hair curled and perfumed, their hoods with lappets of wonderful length... rings on their fingers, and girded belts studded with precious stones of wonderful size, their purses enamelled and gilt with various devices...*"

The unfavourable economic conditions brought about by the sharp reduction in population due to plague led to the Peasants' Revolt of 1381. Rumblings of discontent were being felt in the city three years earlier when the bailiffs were instructed to find out who was responsible for the large assemblies of citizens stirring up debate and discontent who would "*daily cause great disturbances...refusing to submit to justice, and resisting the king's ministers in the execution of their office.*" The archbishop at the time, Simon

Sudbury, was also acting as the chancellor to King Richard II and in November 1380 came up with the idea of a Poll Tax to fund Richard's war with France. The next summer, Kentish rebel leader Wat Tyler led his army of thousands of peasants from Maidstone to London. A branch of this army led by another Man of Kent, John Salos, entered Canterbury on the 10th of June, capturing the city and castle with little resistance. Accounts describe 4000 men entering the Cathedral, leaving Becket's shrine intact but shouting at the monks to elect a new archbishop, Sudbury being a traitor to his people. The Archbishop's Palace was attacked, prisoners from the castle were released and its keeper was forced to swear the Peasants' Oath and burn all of the county's financial and judicial documents in his possession. Damage was also done to the Abbey and enemies loyal to the king were sought out and executed. The next morning several thousand peasants headed off to join Wat Tyler on his march to London. The rebellion culminated a few days later with the beheading of Archbishop Sudbury on Tower Hill in London. The peasants would almost certainly have left Canterbury through the newly built Westgate, ironically having been paid for as a gift to the people by Sudbury, one of the few archbishops of the time to take any real interest in his cathedral city.

More trouble between the clergy and the city occurred in the 1420s. In 1426 a young goldsmith being pursued by the bailiffs for some unknown crime fled to the Cathedral and took refuge behind the iron railings of Archbishop Chichele's tomb (Chichele didn't die until 1443 but had his tomb constructed while still alive). Despite violent attempts to extract him while High Mass was being sung, the Prior ended up sheltering the criminal for forty days, stirring up much resentment among the citizens. In 1427 things came to a head when a dispute flared up over the jurisdiction of a meadow known as the Rosiers, just outside the Westgate. The Priory monks accused the citizens of making hay there, seeing it as Christchurch land, while the citizens accused the Prior of diverting the course of the Stour so as to run past one of his mills. The Archbishop intervened and ruled that the meadow was indeed Priory land but instructed the Prior to return the river to its original course. There were further complaints involving market rivalries, maintenance of the city ditch and jurisdiction over citizens living in the Burgate ward.

A shift from ecclesiastical to civil power was starting to happen across the country, although not as quickly in Canterbury due to the prominence of its religious institutions. St. Augustine's Abbey, lacking anything comparable to the shrine of St. Thomas, no longer exerted much of an influence on city life but was often used by visiting royalty. As well as many kings and Holy Roman Emperors making the pilgrimage to Becket's shrine, there were royal marriages and funerals held at the Cathedral. "Black Prince" Edward of Woodstock, son of Edward III and a national hero for his military exploits in France, asked to be buried there. He was given an elaborate funeral in 1376, his hearse being pulled by twelve black horses. Edward's nephew Henry IV was also buried in the Cathedral in 1413. A couple of years later, on his way back from winning the Battle of Agincourt, his son, Henry V, was welcomed in Canterbury by Archbishop Chichele and his choirboys.

Late Medieval

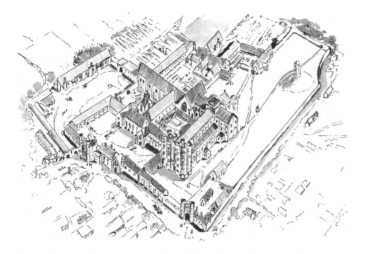

left: St. Augustine's Abbey in its final stages before Henry VIII ordered its destruction in the late 1530s (it wasn't dismantled until 1548); *right*: the 1380 Westgate, funded by Archbishop Sudbury (beheaded in the Peasants' Revolt), with a drawbridge for crossing the extramural branch of the Stour

left: a medieval pilgrims' souvenir, a lead "ampulla" which would have held a tiny amount of "holy water", decorated with an image of martyred Archbishop Becket; *centre*: another mass-produced lead souvenir, a Thomas Becket badge to prove its wearer had pilgrimaged to Canterbury (craftsmen and merchants along Mercery Lane would have been selling these in large numbers); *right*: the shrine of St. Thomas in the Cathedral, covered in gold and jewels and with a removable "lid" on a pulley system – containing his bodily remains, this was built to house the source of the supposed spiritual power which the pilgrims were coming to tap into

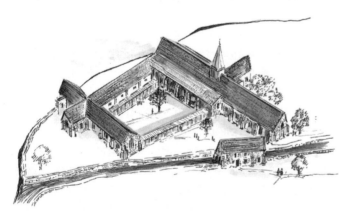

left: The Greyfriars estate would have looked something like this at its most developed, partly built on the small island of Binnewith (or just "With" in some sources) in the Stour. The dormitory in the foreground which bridges one of the sub-branches of the intramural branch of the Stour is all that still stands (in 2003 it was reconsecrated by Anglican Franciscan monks as a chapel); *right*: the striking black flint Blackfriars Gate on St. Peter's Street (the Westgate end of the high street), built in the mid-14th century

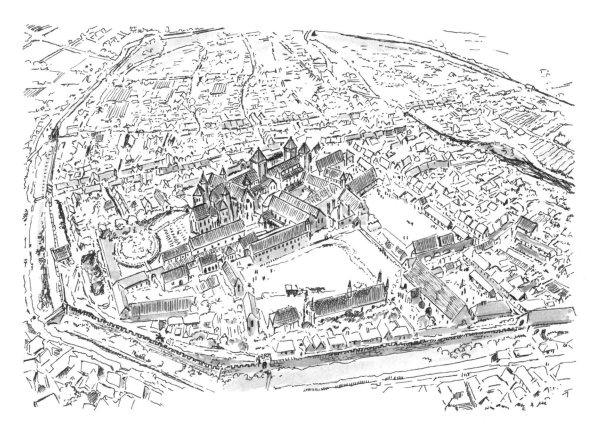

an impression of Canterbury at its medieval pilgrimage peak: Northgate (which actually faces northwest) is at "5 o'clock" here, Burgate at 8, Newingate (St. George's Gate) at 9 and the Ridingate at 10, Worthgate about 11, Londongate at 2, Westgate about 3

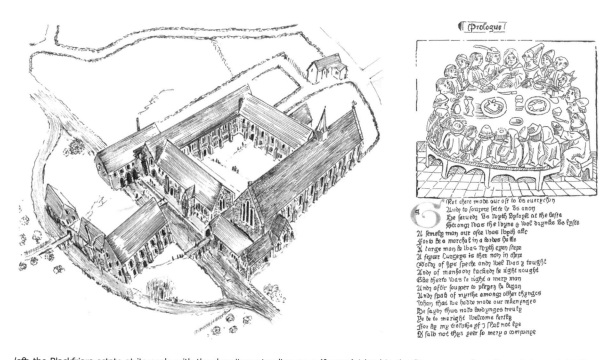

left: the Blackfriars estate at its peak, with the dormitory standing on a (former) island in the Stour, across from the refectory; *right*: the prologue page from William Caxton's 1483 printing of Geoffery Chaucer's *The Canterbury Tales*

14th June 1439 to 26th March 1556

Canterbury's economy was booming in the 1440s, its fortunes improving further when Henry VI issued a new charter in 1448 replacing the two bailiffs with an elected mayor. In 1461 Edward IV issued another charter giving county status to the city "forever" so that the Sheriff of Kent no longer exercised jurisdiction in the city, now effectively a county of its own. As a county required a sheriff, a member of the Burghmoot was elected to this post each year. His roles included delivering the "Fee Farm" £60 lump sum tax to the Crown at Westminster each year as well as some policing functions including the overseeing of public executions. 1497 saw yet another charter, issued by Henry VII, reorganising the governance of the city so that there were more aldermen, the mayor having to be elected each year by the city's freemen from a shortlist of two of these. The idea was to prevent sudden change involving a charismatic leader becoming mayor. The slow and steady continuity that this was intended to provide remained a feature of the ciyt's life for the next few centuries, Canterbury's civic leadership having been notoriously conservative throughout.

Early mayors inherited a history of conflict with the two main religious houses. In 1469 the mayor confronted the Abbey over the ownership of disputed farmland which lay outside the city walls but within the newly created "County of Canterbury". A compromise was reached, suggesting a decline in Church power at the time. In July 1500, the mayor led a group of armed citizens to eject some Christchurch monks from the Rosiers meadow outside the Westgate where they were making hay. This was another dispute based on the new city-county borders and civic-versus-Church jurisdiction. The monks and their servants produced weapons and fought back but the citizens won, leaving the mayor to have to fight a drawn-out legal case in London relating to his actions.

After periods of economic decline and revival, local pride reached a peak in 1480, with new windows installed in the Guildhall and budgets set aside for street cleaning, new uniforms for the city band and St. Thomas's Marching Watch. The latter was an annual pageant involving a military display, sporting contests and a float with boys in tinfoil armour acting out the murder of Becket (involving ingenious automata and pig's blood). Plays were performed near the Westgate by the Guild of Corpus Christi. These "mystery plays" enacted key Biblical stories ("mystery" confusingly referring to the Latin word for "occupation" rather than something mysterious). After Pope Innocent III issued an edict in 1210 outlawing public stage performance by members of the clergy, the various guilds of craftsmen in European towns and cities had taken over this role, each being responsible for one or more stories which formed several complete "cycles". By the 1480s the various guilds, having been around for centuries, began to be formally incorporated and in some cases merged. Each guild had its own patron saint and associated feast day, and members would traditionally attend each other's weddings and funerals. In 1526 a Fellowship of Waits and Minstrels was formed. Before this, the Burghmoot had been employing its own waits (musicians) for some time, using them for weddings, christenings and surprisingly pagan Mayday celebrations.

By 1500 the central tower of the Cathedral had been built up to its present level and the bell "Harry" installed. This had been cast in 1288 and named after the popular Prior Henry of Eastry. The tower subsequently became known as "Bell Harry". This was the last great work to occur on the Cathedral, everything since having been relatively minor adjustments. Around the time that Bell Harry was going up, a new Newingate was being built from the same Maidstone ragstone and with a similar design to the Westgate, its century-old predecessor. Having lost many of its parishioners to plague, the church of St. Mary de Castro was demolished, its parish being absorbed into that of nearby St. Mildred's (its churchyard is now a public green space).

The first stirrings of the Wars of the Roses which were to shake England from 1455 to 1487 were felt in Canterbury in 1450 with a pair of uprisings linked to what's become known as "Jack Cade's rebellion". Unlike the Peasants' Revolt of the previous century, this unrest involved numerous knights, squires and gentlemen taking issue with King Henry VI's advisors and the corruption surrounding his royal dynasty. That February, a local fuller (cleaner of wool) named Thomas Cheyne posed as "Bluebeard the Hermit" and attempted to stir up a rebellion in the city. Failing, he was taken to London, hung, drawn and quartered, his head returned to be displayed over the Westgate as a warning to other potential rebels. Later that year, rebel leader Jack Cade and his 4000 men marched on Canterbury, camping in the land between Harbledown and the Westgate. Mayor Clifton, backed by local opinion, refused to open the gates. The rebels eventually left after doing some damage to some Priory estates outside the walls, marching on to London where they were defeated.

The pivotal Battle of Northampton in 1460 – where Henry VI was captured, leading to a brief change of kings – had repercussions in Canterbury which was divided between Yorkists and (Henry-supporting) Lancastrians. A local landowner called William Pennington had become extremely unpopular with the locals after cutting a ditch blocking their access to the Don Jon Field (below the mound, a popular place for assembly and recreation). It seems that due to his Lancastrian affiliations he was untouchable while Henry reigned, for as soon as the King was dethroned an angry mob beheaded Pennington on the disputed field.

The next year, Yorkist earls gathered across the English Channel in Calais to issue a proclamation condemning the King's government. Yorkist supporters in Canterbury pinned ballads praising the earls to the city gates. Kent was of great political importance at this time because of its geographical position between London and France, Calais being a possession of the English Crown from 1347 to 1558. Later that year Henry VI reclaimed the throne and divisions in Canterbury were made clear when a band of local Yorkists led by the merchant John Bygge rode north to support Henry's rival Edward, while Mayor Nicholas Faunt proudly wore his (Lancastrian) red rose and travelled to London to sit with Henry and his ministers in council. The Mayor paid for his choice of loyalties when Edward retook the throne in 1471. Joining forces with one "Thomas the Bastard", Faunt assembled a large group of Lancastrians (including yeomen and gentry, funded by city finances) to proceed to London to overthrow Edward and reinstate Henry. When they failed, Faunt was brought home to be hung, drawn and quartered just outside the Cathedral's Christchurch Gate, an area then known as the "Bullstake".

An unprecedented rupture occurred in the life of Canterbury in the 1530s when King Henry VIII decided to take on the power of the priories, abbeys and monasteries around the country and to break from Rome, effectively appointing himself the pope in England. In his early reign, Henry had made several trips to the Cathedral, paying tribute at Becket's tomb like any other good Catholic. He would often drop in on his way to or from France, and in July 1520 (the sixth and last Jubilee, celebrating 300 years since the translation of Becket's remains) he even brought his nephew, the Holy Roman Emperor Charles V. That year, the Pope had demanded a full 50% of all shrine offerings, something Henry had grudgingly accepted. Although triggered by the complexities of Henry's marital life, the break from Rome had been brewing for a long time. The common people were starting to realise that their financial contributions to the Church (voluntary or involuntary, direct or indirect) were funding opulence in a distant papal state.

In 1528 Elizabeth Barton, a maidservant from the Kentish village of Aldington, became known locally as a visionary. It seems that she'd been having some kind of seizures and naturally (for the time) interpreted her experiences in religious terms. She was brought to St. Sepulchre's Convent just outside Canterbury where she took a nun's vows. Before long she'd become a national celebrity, the "Holy Maid of Kent", attracting many visitors from afar to hear her visions and prophecies. Her most controversial pronouncement was that Henry VIII would die soon after if he married Anne Boleyn. Either because Henry believed in her prophetic powers or because she was arguably more popular than him at the time, he granted her an audience at least twice. But in the winter of 1533 Elizabeth and two of her supporters from the Priory were arrested and charged with fraud. The implication was that she was being manipulated, her visions politicised as part of a power struggle between Church and Crown. The next April all three of the accused were hung in London, Elizabeth being spared the drawing and quartering inflicted on the other two.

Henry's Lord Chancellor Thomas More had taken an interest in Elizabeth's prophecies and Henry, having come into conflict with him, tried to implicate him in the affair but couldn't find sufficient evidence. More was eventually condemned to death on separate charges and beheaded in 1535. His head was later retrieved from a pike on London Bridge by his daughter Margaret (Meg) who'd married into Canterbury's prominent Roper family. The head was embalmed in spices and placed in a lead box which she kept with her at all times for the rest of her life. It was eventually buried with her in the Roper family vault in St. Dunstan's Church.

In 1534 the Prior and monks of Christchurch, under duress, switched allegiance from the Pope to the King. The actual Dissolution of the Monasteries occurred between 1536 and 1538, confused by the fact that during this time the city council was controlled by Protestant radicals. Although it was a growing force in Europe, Henry's break from Rome was not meant to involve an adoption of Protestantism (rather a kind of independent English Catholicism). The council backed a Protestant printing press and banned St. Thomas's Marching Watch. Through their anti-clerical manoeuvrings they managed to acquire 98 houses and the Abbot's Mill for the city when the Abbey finally surrendered in

July 1538. The Abbey itself was demolished in the years that followed, Henry appropriating the best stone for building elsewhere and the rest being sold locally. Several of the Abbey's outbuildings were spared, though, and converted into a royal residence.

The Greyfriars spent four years under house arrest before finally being suppressed in 1538. The Prior of the Blackfriars sensibly fled. All three friaries became Crown property, eventually being distributed to friends of Henry. John Stone, a principled Whitefriar, refused to sign the required declaration pledging allegiance to the King rather than Rome and for this was hung atop the Dane John Mound, his heart being cut out while he was still alive and his head later displayed above a city gate. Although proud of their Cathedral (the first, largest and most lucrative in the land), the local people had long been involved in bitter disputes with the Abbey and Priory so weren't too upset to see them go. People were more troubled to see the various groups of friars dissolved as they'd lived closer to their ideals of poverty and service, more actively engaging with the local population.

In April 1538, Henry issued a writ accusing Becket of being a traitor against Henry II. As Becket had died in 1170 and wasn't around to plead his case, the King's Attorney General decided that all of the treasures accumulated at his shrine were forfeited and therefore Crown property. That September the shrine was demolished, all of its gold and jewels being carted off to London. The bones of St. Alphege, St. Dunstan and others were dug up and removed and Becket's bones were supposedly burned outside the Cathedral, although this is still disputed. The Priory finally surrendered in March 1539, at which point the Cathedral ceased to operate alongside a monastic institution. A "New Foundation" known as the Dean and Chapter was established to manage Cathedral affairs and many of the monks were recycled into the newly created King's School, by some accounts the first public school in England. There was still an archbishop but rather than being the pope's representative in the kingdom, he was now just another subject of the monarch. Despite all these changes, respect for the Church continued among the populace but the pilgrimages which for more than 350 years had been central to the life of the city (economic and otherwise) were now over.

When Henry died in 1547, the throne was left to his sickly nine-year-old son Edward. The boy-King's councillors (appointed by his father to govern the country until he was eighteen) continued Henry's anti-papist programme during his six-year reign, with chantry chapels in the Cathedral and statues of saints demolished, tapestries destroyed and stained glass removed. Already there was a noticeable blurring of Henry's plan to simply break with Rome and the new Protestant currents which wanted to break from Catholicism altogether. When Edward's half-sister Mary came to power in 1548, this all got reversed, with statues, images and windows being replaced or restored. As a fervent Roman Catholic, Mary had Archbishop Thomas Cranmer placed under house arrest in the Deanery and began fiercely persecuting Protestants. Twenty-three were burned for heresy in Canterbury between July 1555 and January 1556, as well as elderly suspected Protestants being evicted from the almshouses in Harbledown and Northgate. Cranmer was burned at the stake on 21st March 1556 and Roman Catholic cardinal Reginald Pole became the new archbishop, collaborating with Mary in restoring relations with Rome.

26th March 1556 to 22nd April 1649

The persecution of local Protestants continued under Archbishop Pole. Another eighteen were burned at the stake (near Wincheap), bringing the total to forty-one before Mary's death in 1558. This may well have continued had Mary not died then (the last burnings were just two days earlier). Reginald Pole died within twenty-four hours of his queen, at a time of national crisis – Calais, held by the English for over 200 years, was finally recaptured by the French and so the final days of Mary's reign would have seen vagrant soldiers roaming the streets of Canterbury. Despite the persecutions, Mary and Pole had some support in Canterbury, people thinking that their Catholic "counter-reformation" might lead to a revival of pilgrimage and bring back the city's prosperity. St. Thomas's Marching Watch was revived during her reign but the pilgrimages were never reinstated.

Mary's successor, her Protestant half-sister Elizabeth, brought finality to the break with Rome. The tables turned in Canterbury during her reign with several Catholics hung, drawn and quartered on Oaten Hill in 1588. Matthew Parker, her appointed archbishop, had the Archbishop's Palace rebuilt (it had burned down in 1543) and when it was completed in 1565 hosted a "reunion banquet" there to unite the estranged civic and Cathedral authorities. In September 1573, Elizabeth celebrated her 40th birthday with a luxurious twelve-day stay in Canterbury, hosted by Parker, with the Queen staying at St. Augustine's. Much banqueting involving the Cathedral and city authorities, merchants and dignified visitors from abroad was an expensive honour for Canterbury, allowing it to make peace with the Crown, boosting local morale and reassuring itself of its national importance despite the Dissolution of the Monasteries. A couple of years later Elizabeth granted the city the Poor Priests' Hospital (which her father had seized) and an endowment for its workhouse.

A national economic boom saw the woodlands surrounding the city begin to be cut down for ships, houses and charcoal for smelting metal. With the collapse of the pilgrimages, the local economy now hinged on the wool trade, the excellent pastureland around the city being used almost exclusively for sheep. In this, along with London and Norwich, Canterbury now led the way, cushioning the economic blow Henry had inflicted with his Dissolution. Raw wool was being exported to Flanders for weaving alongside woven goods being shipped out of Fordwich. There was so much of this activity that for a while there were even wharves in Canterbury itself, but despite legislation it proved difficult to keep the Stour navigable due to the profusion of mills along the stretch of the river out to Fordwich.

The persecution of Protestants across the Channel which had been going on since 1547 had led some Huguenots and Walloons to seek refuge in England. Mary had ordered all of them out of the country but under Elizabeth they again began to arrive. The Burghmote Court gave them permission to settle in the city and after a major massacre of Huguenots in 1575 a hundred families were allowed to take up residence. Among the refugees were many skilled weavers and craftsmen in silk, providing a major boost to the local economy. The incomers were consequently made so welcome that a stream of them continued to arrive long after the persecutions ended. The Cathedral authorities agreed to let them use the Crypt where they held services in French with congregations reaching 2000.

In 1578 the twenty-four-year old John Lyly, who'd grown up in Sun Street, published *Euphues, or the Anatomy of Wit*, described by some as the first English novel and including a passage in praise of the Cathedral. Involving an ornate style full of antitheses, alliteration and rhetorical questions, it was a major influence on Shakespeare's style. Two years later, the sixteen-year-old King's School graduate Christopher Marlowe, born into a humble family in St. George's Parish, left for Cambridge University where his poetry and plays quickly captured the attention of Elizabethans in high places. This led Marlowe to be given a royal appointment which lasted until his premature death in an East London tavern brawl in 1593. The year before, while visiting his family, the turbulent playwright had ended up in court for fighting a duel with a Cathedral musician in the courtyard of the Chequer of Hope Inn. Like Lyly, Marlowe's writing style can be seen as preparing the way for Shakespeare. In fact, there's an entirely plausible "Marlovian" theory with sane and intelligent adherents which, noting the espionage and intrigue he was now known to have been involved in, proposes that he faked his death, went into exile and wrote most of Shakespeare's plays using his Stratford colleague as a front. Lyly and Marlowe embodied a cultural shift that was afoot: previously, aesthetic concerns (the arts) were associated with priestly/monastic life while material concerns (the crafts) were the domain of the secular population, but things were now changing.

From these times on, there was less and less to distinguish the city from others in England. With a few exceptions, archbishops took no real interest in the place. Some never even set foot in Canterbury. Edmund Grindal brought things to a new low by sending a proxy to be enthroned on his behalf in 1576. The streets at this time were roughly cobbled and there was very little street lighting. Although a time of relative prosperity, day-to-day life was still hard for the masses. Open drains ran into the Stour and many people worked from small, poorly lit houses with rushes on the floor. Chimneys and fireplaces were still a rarity, a central hearth being the norm. There were seasonal festivals to break the monotony of continuous work but the mystery plays of earlier times were no longer the main entertainment. Instead, troupes of strolling players now performed in inn courtyards, their bawdy secular material starting to displace religious stories. As elsewhere in the country, although seemingly not in large numbers, women were being tried in Canterbury's court for witchcraft – there's a record of a "Mother Hudson of St. Mary's Parish" standing trial (but no record of the outcome).

When James VI of Scotland became James I of England in 1603, Canterbury's castle was in a state of ruin. The King gave it to his "Clerk of the Kitchen", Sir Anthony Weldon, and from then on it remained in private hands. A couple of centuries later it was to house the city's water supply but the first movements in this direction began in 1620. Archbishop Abbot attempted to secure a water supply for the city involving a conduit behind St. Andrew's Church (which stood in the middle of the high street) but its awkward position led to the project being scrapped. A few years later a well was dug and a pump installed at the fishmarket in St. Margaret's Street. These improvements were made with plague prevention

Elizabethan

in mind but 1635 saw a particularly bad wave of plague hit the city. Further waterworks were introduced the next year as a result. Also in 1636, local lawyer William Somner published *Antiquities of Canterbury*, the first serious study of the city's already rich history. He'd grown up in Castle Street and apprenticed to his lawyer father, surrounded by court archives full of case histories containing information about the city in medieval times.

The practice continued whereby only freemen were able to trade in the city but as the population grew and every man born of a freeman or married to a freeman's daughter automatically gained his "Freedom of the City", the proportion of freemen within the population grew and so the Members of Parliament elected by the freemen gradually became more representative. Canterbury was able to elect two Members of Parliament but it was also expected to support them financially (being an MP wasn't a paid position back then) which not everyone was happy about.

Although not actually a Roman Catholic, King James was in favour of Roman Catholic "high church" practices that aroused the suspicion and resentment of a lot of his subjects and this was certainly the case in Canterbury. Puritanism was the extreme reaction. The civil war that was to follow was not so much between the Crown and Parliament as between supporters of "high church" practices (typified by James' son Charles I) and "low church" practices (typified by Parliamentarian leader Oliver Cromwell) practices. Canterbury had been strongly anti-papist since Mary's persecutions so unsurprisingly it initially stood with the "low church" dissenters. But this had unintended outcomes.

In 1626 Charles I married (the Roman Catholic) Princess Henrietta of France at the Cathedral, the couple staying at St. Augustine's. Despite the city's anti-papist feelings, the local dignitaries showed their loyalty through hospitality. The King's marriage led him to grant Roman Catholics freedom of worship across the Kingdom. This move was not well received and the fact that the Archbishop of Canterbury at the time, William Laud, was very "high church" and close to the King meant that he too was suspected of being sympathetic to Roman Catholicism. Laud's unpopularity grew when he accused the local Huguenot and Walloon immigrants of not using a literal enough translation of the Book of Common Prayer in their French language services in the Crypt. He disbanded their congregation, scattering them among their various parish churches. The local authorities weren't at all happy with this as it meant that they'd have to support the poor from that community rather than it continuing to support its own poor. Understandably, the immigrants were equally unhappy.

As in London and elsewhere, the people of Canterbury had strong grievances against the King involving taxes and religious interference. The struggle that began the Civil War was an attempt to push the King in the direction of discouraging Roman Catholicism and forcing the bishops to drop "high church" practices. Despite the country being on the verge of war, Charles visited the divided city in 1641 and was again warmly received by the Mayor and council whom he thanked for helping to fund his Scottish military campaign. Meanwhile, Parliament had decided to refortify Canterbury and have it restocked with ammunition.

In 1642 open war erupted between Royalist and Parliamentary forces. Representing the latter, Colonel Sandys entered the city with his men, demanding the keys to the Cathedral. When they found arms and a store of gunpowder put there by Royalists, a destructive rampage followed with shots even fired at the statue of Jesus above the Cathedral's Christchurch Gate. Expressing strong anti-papist feelings, Sandys' men tore up all the prayer books in the Cathedral. Stained glass was smashed, statues destroyed and valuable ornaments stolen. Although they'd initially been supportive of the anti-Royalist forces, many locals were now horrified. The next few years are a confusing episode in the history of the city as all we have to go on now are rival pamphlets from that time. It seems that people thought it would all end quickly with the King giving in to Parliament's demands. But it was to be another seven years before things reached an unthinkable conclusion, during which time ideas about monarchy, the Church and what was being fought for gradually changed. Puritans tended to become more extreme and hatred of the bishops grew, people believing that they were conspiring with the King to bring about a return to Roman Catholicism.

In January 1645 Archbishop Laud was executed at the Tower of London for treason after four years in prison for supporting the Royalist cause. There was to be no Archbishop of Canterbury for the next fifteen years. The same year that Laud was executed, an almost 200-year old market cross was removed by the Puritan mayor John Pollen from the area known as the Bullstake, just outside Christchurch gate. Inside the gate, Parliamentarian troops were using the Cathedral Precincts as barracks and stabling their horses in the buildings. Puritan extremism had reached a peak and this caused a backlash. Some people were holding private Church-of-England-style services in secret, risking arrest and possibly imprisonment. Things reached a climax in 1647 when the Mayor banned Christmas and decreed that shops were to stay open for business. When he shut down a Christmas service at St. Andrew's Church and accused the congregation of unlawful assembly, he triggered a riot. The dozen shops that were open as instructed had their wares destroyed, "*For God, King Charles and Kent*" being the rebel cry. The prisons were emptied, ale was distributed freely and football was played in the high street. This popular uprising had nothing to do with supporting "high church" practices, people just wanted to celebrate Christmas – they'd had enough! A petition was circulated and a group of men marched to London with it, this being how the "Insurrection of Kent" began, the people turning on the Parliamentary forces they'd initially supported. Scattered by Parliamentary troops near Maidstone, though, their insurrection failed.

The next May, trials were held for the supposed leaders of the Christmas revolt. The jurors' unwillingness to convict them played a pivotal role in sparking off the next phase of the Civil War. This culminated in General Fairfax leading 7000 Parliamentarian troops into Kent, defeating the Royalists at Maidstone on 1st June 1648. Canterbury was the last stronghold of revolt, but with only 1300 men to defend it the city surrendered. Gates were pulled off their hinges and burned, and a breach in the walls fifty yards wide was made just south of the Westgate so that the city could no longer defend itself. On 30th January 1649 Charles I was beheaded in London with his chaplain William Juxon at his side and a period known as the Commonwealth began.

22nd April 1649 to 25th June 1723

The Commonwealth lasted for eleven years. Early on, Parliamentarian leader and now "Lord Protector of the Commonwealth" Oliver Cromwell visited Canterbury and received an official welcome and banquet. Despite the city having made its peace with the new regime, the demolition of the Cathedral was seriously discussed in Parliament a year later due to Canterbury's Royalist sympathies during the Civil War! The effects of Puritanism were cushioned in Canterbury by a succession of more moderate mayors, with a kind of religious tolerance gradually displacing strict Puritanism. The French-speaking immigrants were left alone and Church-of-England-style services continued privately. By the time Cromwell died in 1658, people had really had enough, welcoming a return to life as before, with bawdy plays being performed in inn courtyards, Mayday festivities, dancing and a more colourful style of religion.

After Cromwell's son failed to maintain control of the Commonwealth, the Monarchy was restored in June 1660. Charles' son (soon to be Charles II) returned from exile, spending a night at St. Augustine's. He attended a service at the Cathedral before journeying on to London and was given a gold tankard, a Bible and a copy of William Somner's *Antiquities of Canterbury*. That September, William Juxon, the royal chaplain who'd been with Charles I at his beheading, was enthroned at Canterbury. Enough locals were glad to have an archbishop again after fifteen years without one that they were able to line the streets from the Westgate to the Cathedral to cheer him in. The first archbishop to have a positive relations with Canterbury for some time, he made personal contributions not just to Cathedral repairs but also to the upkeep of the city's walls and gates. He also struggled to find suitable priests for the many local parish churches which had seemingly become infested with charlatans during the Commonwealth.

William Somner's brother John was a tailor who operated from premises beside the Cathedral's Christchurch Gate and it was he who had persuaded Archbishop Juxon to replace the city gates. He went further and spent £400 on having a market house erected in the Bullstake. It had an open lower storey, with the upstairs intended for public meetings and storing grain. In 1658 Sir Edward Hales took possession of the Abbey grounds after the death of Lady Wootton, who'd been living there since the 1626 death of her husband Edward, Lord Wootton of Marley. Hales and his son weren't interested in conserving what was left of the Abbey, instead dismantling many of the buildings and moving the stone up to St. Stephen's Parish to build a huge mansion which became known as Hales Place. Development began to speed up at this point. A 1663 drawing of the city shows that dwellings within the city walls were relatively sparse at the time with the Cathedral and castle standing out. A similar 1735 drawing shows that much building had occurred in the intervening decades.

Canterbury saw several waves of plague in the 1660s, the unusually hot summer of 1665 bringing with it a particularly bad epidemic. As well as the plague, the people had a possible war with the Dutch to fear. Although this never happened, Dutch warships did sail up the Medway and seize nearby Sheerness. The city's walls and gates were strengthened in response to this, still offering some meaningful protection despite the growing use of gunpowder.

By 1676 silk weaving had become the city's dominant commercial activity, outstripping wool. A thousand looms employed 2700 natives and immigrants. A decade later another wave of religious intolerance across the Channel brought 1500 more Huguenot refugees, including skilled weavers who boosted the local textile industry. However, some in the city complained that they were mostly poor, sickly or elderly. Canterbury's relative prosperity post-Dissolution was partly due to its fortuitious location close to the source of skilled immigrants, but also centuries of practice at providing hospitality meant that they largely felt welcome to settle in the city despite some local reservations.

The English wool boom soon came to an end, though, with the number of local looms down to 200 by 1710, operated by a workforce of just 540. There was a brief reversal of Canterbury's fortunes when a local master weaver developed a technique for producing a mixed silk-cotton fabric which became fashionable around the country for a few years, known as "Canterbury muslin". A mill was built on the Stour which employed fifty women and children to produce this desirable commodity for about a decade. But with the decline in textiles, by the end of the 17th century hops had become the dominant local commodity. Rather than being exclusively used for sheep pasture, the open land around the city began to be cultivated for hops, orchards and mixed farming. After centuries at the centre of power struggles and pilgrimages, Canterbury had been reduced to a fairly insignificant market town.

In 1717 the *Kentish Post* (later to become the *Kentish Gazette*) was first published in Canterbury, by some accounts making it the second oldest newspaper in country. It was printed weekly by Thomas Reeve in Castle Street, initially as *The Kentish Post: or the Canterbury News-Letter*. The *Post* was quick to adopt the serialisation of novels, beginning with Daniel Defoe's scandalous *Moll Flanders* in 1722-23. Another hit with the locals was the arrival of public street lighting in 1687: one big oil lamp installed at the Bullstake.

Charles II was popular in the city but his brother and successor James II much less so. Fears of a return to Catholicism and Mary-style persecution led to James being pushed into exile only three years into his reign with his Protestant daughter Mary and her husband William of Orange being invited in from Holland to take the throne in 1688. They arrived at Torbay in Devon so didn't pass through Canterbury on their way to London, but William was welcomed to the city in both 1690 and 1694. A street near the Cathedral Precincts was eventually renamed "Prince of Orange Lane" (after a tavern of that name), later to be abbreviated to simply "Orange Street". 1714 saw the beginning of the Georgian dynasty, a line of German-speaking kings who, when travelling abroad, tended to sail from Harwich to their native Hanover, so after many centuries of regular royal visits Canterbury witnessed very few in the 18th century.

From the 16th century Reformation to mid-19th century, archbishops were not buried or even commemorated in the Cathedral. They spent very little time in Canterbury, enthronement by proxy becoming the norm. A great storm in 1703 badly damaged the Arundel (northwest) Tower of the Cathedral and – an indication of the diminishing importance of the geographical site of the Mother Church in England – it wasn't rebuilt until the 1830s.

left: a possible portrait of Christopher Marlowe, based on a painting found at Corpus Christi College, Cambridge in 1953, dating from around the time he would have studied there (of course, he may have looked nothing like this); *right*: the medieval pilgrims inns with their courtyards as they would have looked around 1520, the Chequer of Hope being most prominent (although the pilgrimages had been suppressed by his time, Christopher Marlowe appeared in court in 1592 for fighting a duel in the Chequer's courtyard when visiting his family that year)

left: the 17th century market house that stood in the Bullstake area outside the Cathedral's Christchurch Gate; *right*: the ruins of Ethelbert's Tower, demolished for building materials after Henry VIII's Dissolution of the Monasteries – built under the direction of Abbot Scotland (1073-87), it had been one of the finest examples of late Norman architecture in Britain

left: the crest of the Dean and Chapter set up to manage the Cathedral after the Dissolution, also the crest of The King's School which was set up at the same time (the "iX" stands for "Jesus Christ" in Greek); *right*: Hales Place, a mansion built by the Hales family in St. Stephens Parish, Hackington using stone taken from the Abbey site after its suppression

the Newingate as it would have looked when it was built in the late 15th century – modelled on the earlier Westgate (on the opposite side of the city walls), it later became known as St. George's Gate, named for the parish that it lay within

left: the Worthgate, with its original Roman arch – in Roman times this would have been where the road from Lympne entered the city; *right*: the Northgate, with the chancel of St. Mary's above it (Mary was easily the city's most widely celebrated saint pre-Becket)

left: the Burgate, where the road from Sandwich entered the city – it faced the Abbey's Cemetery Gate (built in the late 14th century in the style of the main Fyndon Gate) which has survived it; *right*: the Cathedral's distinctive "Bell Harry" tower, its last major work, completed around 1500

25th June 1723 to 5th August 1782

After all the turbulence of previous centuries, the 18th was a relatively quiet one for Canterbury. The inside of the city walls was being developed and the suburbs were expanding. There was still a huge wealth gap between rich and poor, life being particularly hard for the latter. Children worked as soon as they were old enough. Workhouses were starting to appear around the city with everyone from infants to the elderly working long hours. Prostitution thrived. Gin, cheap and widely available, became very popular.

In 1728 a Board of Governors was appointed to look after sixteen "Blue Coat Boys" at a school based in the old Poor Priests' Hospital. The school had been founded in 1574, funded by endowments left by Archdeacon Langton in the 13th century, and would eventually evolve into the Simon Langton Grammar School. There was a windmill on top of the Dane John Mound at this time (it was advertised for sale the next year) which must have been built after the Civil War, as the mound had then been used as a gun platform. In 1730 Jews returned to Canterbury. Oliver Cromwell had begun to tolerate a Jewish presence in London from 1656, needing their moneylending services to finance his Irish campaign. In 1762 a new synagogue was built in the St. Dunstan's area where the Jewish community had begun to settle. Also in 1730, Freemasons began to operate in the city, initially meeting at the Red Lion Inn (the "talking shop" next to the Guildhall). A long poem appeared in the local press condemning them and a couple of years later a woman fell through the ceiling attempting to listen in on one of their secret meetings from the room above. After Parliament passed the Act of Uniformity in 1662, various types of "nonconformist" Protestant sects began to spring up outside the Church of England – Baptists, Quakers, etc. Methodism took root in Canterbury in the mid- 18th century, its founder John Wesley visiting in August 1764 to open a chapel in King Street, a curious twelve-sided structure which became known locally as the "Pepper Pot".

In 1750 a Mr. Francis Whitfield built assembly rooms at the corner of St. Margaret's Street and the high street. This was used for concerts, balls and large, fashionable evening parties. The upstairs of the market house outside the Cathedral gate began to be used for plays, sometimes after a day's cricket at Bourne Paddock in nearby Bishopsbourne (this was a pioneering venue for cricket, the third earliest "First Class" match of all time being played there in 1772 according to some sources). The link between cricket and theatre in Canterbury was to continue into the 21st century. 1754 saw cisterns installed in the towers on either side of St. George's Gate (originally known as the Newingate), finally supplying decent water to all of the city's markets and the Guildhall.

In 1766 George III granted Canterbury the liberty of a toll-free market to be held every Wednesday for buying and selling hops, allowing the city to become the centre of the East Kent hop trade. A 1768 map shows hop fields planted right up to the city walls and in the land behind the houses lining St. Dunstan's Street and London Road. Not surprisingly, brewing also became a significant part of the local economy, the appropriately named Alfred Beer & Company brewery operating within the Abbey grounds from 1770. Around this time 313 new freemen were incorporated to represent the city's brewing interests.

A major event in the cultural life of Canterbury took place in July 1765 when an eight-year-old Austrian boy called Wolfgang Mozart performed at the Guildhall. He was on his way to play for the royal court in London and to meet Bach there. His family stuck around for a few days, staying near Bishopsbourne. In 1768 the *Kentish Post* suddenly found itself with a rival when local stationer James Simmons founded the *Kentish Gazette*. After a four-week trade war, the proprietors of the *Post* were forced to merge with the *Gazette*. In 1769 the Canterbury Historical Society for the Cultivation of Useful Knowledge was founded, meeting weekly at the Guildhall Tavern. Bathing in the chalybeate springs at St. Radigund's (the "cold bath" just outside the city wall) became popular with a certain sector of local society at this time. Widening the scope of his commercial ventures, James Simmons leased the estate of St. Radigund's and expanded and improved the bathing facilities. For the lower strata of local citizenry, the post-harvest Michaelmas Hiring Fair still providing several days holiday each late September within the Cathedral Precincts.

Extreme flash flooding in January 1776 resulted in the Stour bursting its banks and several people drowning near the Westgate. Three years later, a secular/popular music scene sprung up in the city with the founding of the Canterbury Catch Club. "Catches" were popular songs, sometimes bawdy, sometimes patriotic. Membership was by subscription, and each Wednesday from October to May members would gather in the Prince of Orange tavern. There they would listen to a concert played by a hired orchestra while smoking long clay pipes, drinking ale and eating meat pies, then enjoy a lively singalong involving catches and "glees". A St. Cecilia's Dinner was held each November in honour of the patron saint of music.

Creation and destruction were occurring in roughly equal measures locally. From 1770 to 1792, the outer walls of the castle were demolished, its ditch filled in and houses built on the land, and in the summer of 1781 the Burgate was partly demolished. But a new Corn and Hop Exchange was built in St. George's Street and a new fishmarket established. The area outside the Cathedral gate was still known as the Bullstake at this time, as it had been used for baiting bulls with dogs for centuries, a practice believed to make their meat more tender. The Burghmote had even made baiting before slaughter a legal requirement to sell the meat, despite the occasional escaped bull having wreaked havoc down Mercery Lane! Stretching from the Bullstake to Butchery Lane (formerly Sunwine's Lane and Angel Lane) was a twice weekly poultry market which went back to at least the 14th century.

No 18th century archbishops lived in the city. At most they would pay occasional flying visits. The much less powerful Dean was now responsible for the life of the Cathedral. In 1753, Archbishop Herring sent the Dean a contemptuous letter informing him that the King of Sardinia was interested in St. Anselm's remains. It included a rant against 11th-12th century Archbishop Anselm as a "*rebel to his King, a Slave to Popedom & an Enemy to the married Clergy... I should make no Conscience of palming on the Simpletons any other old Bishop with the Name of Anselm.*" The break with Rome was well and truly established.

Georgian

5th August 1782 to 15th September 1829

The late Georgian era saw the turning point where people in Canterbury stopped having to worry about dying of plague or being hung, drawn and quartered for their religious beliefs. By the 1780s, thieves were more likely to be publicly whipped or transported to the colonies than hung. Generally things were becoming more "civilised", as the prosperous retired people who started taking up residence outside the city gates in large Georgian houses might have told themselves. In 1790 a "dispensary" was founded alongside the Poor Priests' Hospital, an early form of health centre. Examining poor patients, a Dr. Rigden noted that the high water table, lack of piped water, cemeteries within the city walls and back street abattoirs were causing the poor residents of the damp half of the town to live on average ten years less than everyone else.

The traditional Michaelmas Fair (the last of four annual fairs to survive) was last held within the Cathedral Precincts in 1826. It was discontinued because of excessive drunkenness and trading disputes. Local artist Thomas Sidney Cooper disapprovingly described how a "hub-ub caused by swinging boats, merry-go-rounds and other amusements, the shouting, kissing and screaming of the crowds, all took place under the lofty spires and pinnacles of Christ Church Cathedral". The same year, St. Augustine's Brewery was established in the Abbey grounds, the brewery gardens staging fireworks, acrobatics and hot air balloon ascents.

From 1794 military barracks were being constructed just beyond the Northgate. Initially housing 1500 soldiers, and eventually 5000, these had a major impact on the life of the city, its civilian population being only about 12,000. Troops were being stationed close to the coast due to the threat of invasion by Napoleon's forces. In 1803 a chain of beacons was established between Canterbury and the coast as a way to signal such an emergency. Local trouble typical of that time was reported in August 1810 when members of the 2nd Dragoon Guards attempted to seize a suspected Navy deserter in Ruttington Lane, two of them receiving serious knife wounds.

A series of 18th century "Turnpike Acts" funded the surfacing of roads between major towns and cities, revolutionising transport in England. Because of its key geographical location, this helped put Canterbury back on the map. By 1800, the city had sixty-nine taverns, eight hotels and three major coaching inns, as well as many lesser ones. Six coaches plus royal and foreign mail coaches ran ninety-one weekly journeys. It took six hours to London and tickets could be purchased for a journey to Paris, including the sea voyage. A new kind of pilgrim began to show up around this time – the "antiquarian". The origins of modern tourism are to be found in this kind of secular interest in history, with guidebooks starting to be published for the more historically relevant towns and cities. By 1825, hop fields, orchards and market gardens were surrounding the city. Along with hospitality and agriculture, tanning and brewing began to replace the now almost-dead textile industry (cheap and desirable new fabrics were now being imported from various parts of the British Empire).

Despite the rise of antiquarianism, Canterbury had yet to realise the value of its historic buildings. Between 1782 and 1791 the Ridingate, still including arches from the Roman original, was taken down. A Commission for Paving, Lighting and Watching was established in 1787 and decided to remove anything protruding into the street by more than a foot. St. Andrew's (in the middle of the high street) and the tower of All Saints Church both had to go, along with the 400-year-old black flint Blackfriars Gate on St. Peter's Street and countless minor architectural oddities that gave the city its distinctive character. In 1791 the Worthgate was torn down. In 1801-02 St. George's Gate was demolished, a process so costly that the similarly structured Westgate was spared in 1824 when its demolition was proposed in order to ease the movement of hop wagons. The Red Lion Inn next to the Guildhall, purchased by the city in 1408, was pulled down for the creation of Guildhall Street in 1806. 1817 saw the demolition of the upper storey of the castle (the part with all the best stone carving). The last remaining tower in the Abbey grounds was taken down in 1822 and by that year the Burgate was also gone.

By the end of the 18th century, Georgian buildings were replacing Elizabethan ones, or at least Georgian fronts were being applied to them, covering up their half-timbered work which seemed old, clumsy and an embarrassment to the "progress"-minded authorities of the time. The Commission for Paving, Lighting and Watching decided to install 150 iron lampstands and 240 oil lamps, completely changing the life of the city at night. In 1790 the old Market House outside the Cathedral Gate was demolished due to rotting timbers, a large open market house with an oval roof being built to replace it. From this time, the area that had been called the Bullstake came to be known as the Buttermarket. A Mrs. Sarah Baker had been successfully running a theatre in the upstairs of the old market house for some years, so she had a new, large and elegant theatre built on Prince of Orange Lane. Between 1790 and 1803 the area previously known as the Don Jon (or Dane John) field, long used by locals for recreation, was landscaped and turned into the "Dane John Gardens" by entrepreneur (and, by this time, former mayor and alderman) James Simmons, incorporating part of the Roman bank, the medieval wall-walk and the mound. The Dane John Mound was significantly reshaped into a conical form with ascending spiral paths, raising its overall height and creating a pleasing view over the city but forever changing this mysterious, ancient earthwork.

In 1792 an unusually tall structure was built beside the Stour in St. Radigund's, originally intended to be a granary, but then converted into a mill by James Simmons. The next year, the first Kent and Canterbury Hospital was opened in Longport. Nearby, adjacent to the city jail which had replaced the one above the Westgate, a new courthouse was built in 1808 with prominent, curious heraldry involving a fasces (axe-and-bundle-of-sticks symbol that gave its name to fascism) and an ancient Phrygian "liberty cap" (as worn by freed Roman slaves) on a pole. A significant figure in the Commission for Paving, James Simmons had big ideas for Canterbury, putting forward a plan around 1800 to turn the Stour into a canal between Canterbury and the sea so that shipping could reach the city and thereby boost its economy. This idea was given serious consideration but then dropped when the Canterbury-Whitstable Railway started to be planned in 1825. Work began on the railway in 1828. Around the same time, the remains of the castle keep were sold to the Gas, Light and Coke Company who set up the first truly city-wide water supply, installing a large tank high up in the keep and pumping river water into it.

Georgian

upper-left: the banner atop the early 18th century *Kentish Post*, Canterbury's first newspaper, later bought out by James Simmons' rival *Kentish Gazette*; *lower-left*: the oval-roofed structure which replaced the old market house outside the Cathedral gate in 1790; *right*: the Cathedral's Christchurch Gate as it would have looked at this time (based on a painting by Thomas Sidney Cooper)

left: the Dane John Gardens, the ancient mound having been landscaped into a more conical shape, brainchild of stationer, alderman, mayor, newspaper magnate and banker James Simmons at the end of the 18th century; *right*: the medieval Ridingate as it would have looked at the end of the 18th century, in urgent need of repair and soon to be demolished

left: based on the etching *The Freemasons Surprized*, this shows the chaos that ensued when a woman eavesdropping from the room above the early Canterbury masonic lodge's meeting place in The Red Lion fell through its ceiling; *right*: the Wesleyan (Methodist) "Pepper Pot" chapel opened in King Street in 1765

left: All Saints Church in the High Street as it would have looked shortly before its tower was demolished in 1769 as part of a road-widening scheme (the church was rebuilt in 1828 and then demolished in 1937); *right*: Wincheap Gate (built into the city walls not far from the then defunct Worthgate after it was blocked in 1548) as it would have looked shortly before its demolition in the late 18th century

left: The Norman castle in ruins after its upper storey was demolished (1817); *right*: the curious "fasces-and-liberty-cap" heraldry on the facade of the city's courthouse built in Longport in 1808 adjacent to the city's prison, HMP St. Augustine's

Canterbury's 18th century music scene in the form of the "Catch Club", meeting for a weekly Wednesday concert and singalong above the Prince of Orange tavern, the first of several meeting places in its 86-year history

15th September 1829 to 2nd April 1867

The Canterbury-Whitstable railway was completed in 1830. Connecting Canterbury to Whitstable harbour meant a link with shipping, a boon to the city's economy. As well as freight, the train carried passengers, making it the second passenger railway line in the world (passing through the world's first railway tunnel). The opening day was a major occasion, involving local dignitaries and their wives in their finest clothes being transported to Whitstable and back, with a feast at either end. In early 1846 the Canterbury-Ashford line was opened, connecting to the Whitstable line at a new station near the original one. The construction of Station Road off St. Dunstan's Street involved demolishing the city's 1762 synagogue, so a new one was built in King Street – in Egyptian temple style, following the "Egyptomania" fashion of the times. In 1860 a railway line was opened between London and Dover via Canterbury's new East station, built a stone's throw from the Dane John Mound. Unfortunately, the station's construction involved the complete destruction of another of the ancient mounds.

The autumn of 1830 saw farmworkers rioting in the nearby countryside, resisting the introduction of machinery which threatened their livelihoods. Troops were called in to patrol nearby villages and on some nights distant corn stacks and farm buildings could be seen burning from atop the Dane John Mound. Meanwhile, what was left of cloth manufacturing in the city was rapidly dying out due to industrialisation in the north of England as well as cheap imports from India. In general, the distinctive character of Canterbury was being swiftly eroded by fast-paced improvements in communication, travel and the beginnings of mass production in England.

The population in 1841 was recorded as approximately 15,000. Streets were still mostly cobbled and, despite some centralisation of a water supply, women were still seen weaving in and out of traffic carrying buckets. The historical heritage of the city finally began to be widely recognised as something valuable and worthy of preservation, but things were slow to change, with the demolition of the Westgate again being seriously discussed in 1859 so that the elephant-drawn carriages from Wombwell's Circus could enter the city! This was stopped by a single vote, cast by the mayor.

In 1860 Sarah Baker's theatre in Orange Street was replaced by a new "Theatre Royal" in Guildhall Street, funded by the extremely successful local painter Thomas Sidney Cooper. Cooper had begun his career as a scenery painter at the original theatre. He was a personal friend of Charles Dickens, who came to read from his *David Copperfield* at the Theatre Royal in 1861. The local connection between cricket and theatre continued, with the "Old Stagers" founded in 1842 (eventually to become the oldest surviving amateur dramatics company in the world) and performing at the popular annual Cricket Week. Kent County Cricket Club began life as the Old Beverly Cricket Club near Hackington, moving out to the St. Lawrence ground south of the city in 1857. Unusually, this ground was built with a lime tree inside its boundary, with a special rule introduced to deal with balls which hit the tree. Numerous private day schools were springing up around the city, boosting local literacy (believed to be less than 50% in 1800). In 1855 St. Edmund's, a private school founded in Yorkshire in 1749, moved into an imposing

Gothic-style building at the top of St. Thomas' Hill, visible from the Westgate.

Civic-Cathedral relations were again beginning to deteriorate. The appointment of the ultra-conservative Archbishop William Howley in 1828 led to a wave of anti-Church feeling. He was the last Archbishop of Canterbury to wear a wig, regularly spoke out against all forms of democratic or political reform and was strongly opposed to religious tolerance. The similarity of his views and those of the corrupt and oligarchic city council meant he was intensely unpopular locally. That he sent a proxy to Canterbury to be enthroned didn't help. When he finally visited the city in August 1832, his official welcome almost turned into a riot, with rotten fruit, vegetables, eggs and even a dead cat thrown into his carriage. His reception at the Guildhall was cancelled and he spent the night holed up in the Cathedral before fleeing back to Lambeth the next day, never to return.

The Abbey grounds were now being used for grazing pigs. A tavern occupied part of the site, its beer garden used for cock fighting, dancing and raucous music. In 1843 a visitor from London complained about the desecration of this once sacred site in a church newspaper, leading a wealthy MP for Maidstone to purchase the entire site so that it could be used to house a new missionary college. This opened in 1848, sending missionaries out around the world (rather like in the early days of the Abbey) during the peak years of the British Empire.

The backdrop of elite corruption and a largely unrepresented rural populace led to an extraordinary sequence of events in the early 1830s. John Nichols Thom, a tall, bearded Cornishman dressing in exotic clothes, showed up in the area claiming to be "Sir William Courtenay", a Knight of Malta, seeking to "*take the burden of taxation from the shoulders of the poor and industrious classes and fix it on those of the rich*". Standing for office, this charismatic mystery man managed to get 375 out of 2000 votes. His local following grew, assisted by public speeches and a weekly broadsheet, *The Lion*, which he wrote and distributed (this was described as a "*strange mix of biblical quotations, dogmatic assertions, scurrilous accusations and occasional flashes of common sense*"). Accused in court of giving false evidence in a trial, he was sent to Barming asylum near Maidstone for four years. Eventually, his father and wife succeeded in having him released and he was allowed to live at the home of a supporter in Boughton-under-Blean.

The "Poor Law" had been passed while Thom was in Barming, so conditions were now favourable for further campaigning. By 1838, he had a band of followers riding around the nearby villages carrying pistols, swords, a bugle and displaying a standard (a red lion on a blue background). According to some reports he was claiming to be a reincarnation of Christ. Alarmed locals complained and the High Constable of Boughton was sent to issue a warrant. When the Constable arrived at the farm where he was staying, Thom opened fire and killed him. He and his followers then took refuge in nearby woods and troops were called in the next day. During the ensuing "Battle of Bossenden Woods", Thom killed a young lieutenant before soldiers opened fire killing him and seven of his followers. By some accounts, this was the last "battle" involving troops on British soil.

Victorian

2nd April 1867 to 4th March 1897

By 1871 Canterbury's population had increased to about 21,000. An improved mains water supply had been built in Wincheap in 1869, improving public health, but extreme poverty meant that many locals were more interested in drinking gin (still an epidemic despite 18th century legislation to control its consumption). Stour Street was a slum, with seven pubs selling cheap gin, some only yards apart. More prosperous citizens were able to concern themselves with such things as the preservation of the city's historical heritage. By the 1880's the conservation movement was holding major demonstrations with both men and women wearing primroses as an indication of their support. The worst of the destruction was now over, and the unpopular and unattractive 1799 replacement for the old Ridingate was removed in 1883, replaced by a simple iron bridge allowing people to walk along the city wall to the Dane John Gardens. The presence of extensive barracks along the Sturry Road continued to affect the life of the city, particularly in the Northgate district which, due to its proximity, had become a red light district by the 1890s, full of taverns and brothels. Fights broke out so regularly that policemen patrolled the area in pairs. Queen Alexandra's Music Hall, specialising in "low class entertainment", became known locally as the "Penny Theatre" (still operating under this name in 2017). By 1890 cobbles were no longer the norm on the city's streets and bicycles began to appear – in February 1896, complaints were made about "lady bicyclists" using the Cathedral Precincts as a practice area, narrowly missing elderly pedestrians.

By 1880, Freemasonry was well established in Canterbury with three lodges operating (United Industrious no. 31, St. Augustine's no. 972 and Royal Military no. 1149). Collectively, they bought no. 38 St. Peter's Street and built a Masonic Temple there, eventually to be used by numerous other lodges. One such lodge later established was for former pupils of Simon Langton Grammar School, opened on the site of the old Whitefriars estate in 1881, a continuation of the Bluecoat School for Boys established with endowments left by the 13th century Archdeacon (and turncoat) brother of Archbishop Stephen Langton. The next year, local artist and national celebrity Thomas Sidney Cooper opened an art school adjacent to his mother's old house in St. Peter's Street where he'd grown up with numerous siblings in extreme poverty. Cooper had acquired a fortune from his popular paintings of rural and agricultural life, popular with the aristocracy and royalty. He was known locally for his philanthropy, having funded a number of almshouses in Chantry Lane. His father-in-law William Cannon was the leaseholder of the Abbot's Mill (St. Radigund's) until his death in 1875, at which time Cooper stepped in to help Cannon's widow by buying the mill, modernising it, leasing it to tenants and eventually selling it to a Mr. Denne in 1896.

Canterbury's awareness of its own history was considerably expanded in 1871 when engineer James Pilbrow published his findings concerning the layout of the Roman city. This wasn't formal archaeology, rather the result of an interested amateur who'd gained access to subterranean layers of the city via his involvement in the digging of trenches for the new water system. Before this, Canterbury was only vaguely aware of its Roman past, very few details being available. Twenty years later, when the oval-roofed market house was removed from the Buttermarket, the city again looked to its past and decided to honour its greatest literary figure by replacing it with a memorial to Christopher Marlowe. This was unveiled in September 1891 by the celebrated stage actor Henry Irving. Unfortunately, Victorian morality didn't approve of the design – a topless muse holding a lyre. In fact, it didn't approve of Marlowe himself, a supposed atheist who'd died disgracefully. The statue was described as "a voluptuous and frivolous maid" and "an affront to Christianity", and it was eventually removed from its position outside the Christchurch Gate to a site in King Street and then on to the Dane John Gardens.

In 1880, and not for the first time, Canterbury came under the scrutiny of the Royal Commission for its notoriously corrupt election practices. The two main parties, Liberal and Conservative, were found equally guilty and wealthy local former MP Henry Munro-Butler-Johnstone was blamed for "debasing public morality" (but not actually charged). Only freemen could vote and as there were less than 2000 of them, outright bribery had become the norm. As it had been once before in 1853, the city's Parliamentary borough was disenfranchised, leaving Canterbury unrepresented for over five years. At the end of this period, the once busy port of Fordwich was deprived of its corporate status. The Stour had long since silted up so all significant river trade had ceased. The next year, 1887, the highly conservative city council voted against Canterbury being connected to the new telephone network, earning local condemnation for voting "against progress and the telephone". Fifteen years earlier, the Cathedral authorities had shown a similar conservative stubbornness: with the growing importance of rail travel they were requested to keep their clocks set to Greenwich Mean Time but refused, preferring to stick with their own local "Cathedral time".

The Cathedral narrowly avoided a major catastrophe in 1872 when workmen on the roof knocked over a charcoal brazier. There was significant fire damage but it was contained by firemen, cheered on by crowds. For months afterwards, services were packed, local people having a newfound appreciation for their cathedral. 1883 saw the enthronement of Archbishop Edward Benson who had his own strong opinions about how a cathedral should operate, having published a book on this topic. He took "a medieval bishop's interest in the city" and initiated the building of a new Archbishop's Palace, but died in 1896, never seeing the completion of this work. It was completed during the tenure of his successor, Frederick Temple, the first archbishop to live in Canterbury in many years. An excavation of the Cathedral's East Crypt in 1888 found some bones on the site where Thomas Becket had been interred between 1170 and 1220. Skull damage suggested that these might in fact be Becket's bones (widely believed to have been translated to the new shrine in 1220 and then destroyed by burning on Henry VIII's orders in 1538). A debate about the fate of Becket's bones then began, one which has continued into the 21st century.

Some charred bones from the same era were dug up by workmen in the Wincheap area in 1896, those of Protestant victims of "Bloody" Mary's religious persecutions in the 1550s. This led to the erection of the Martyrs' Field Memorial three years later.

Victorian

4th March 1897 to 7th January 1921

By the turn of the century the local population was about 23,000, growing slowly over the next couple of decades while a lot of similarly sized English towns' populations were exploding into the hundreds of thousands. The biggest occupational sector was now domestic service with wealthy families living in large houses along the London Road, Old and New Dover Roads. The working people enjoyed a rollerskating rink (supposedly the country's largest) built in Rhodhaus Town in 1909, while not far round the outside of the city wall, boisterous crowds gathered for music and dancing at the annual fair held at the old cattle market. This had taken over from the annual fairs held within the Cathedral Precincts until 1826 and featured the popular Crown and Anchor Dancing Booth, described in a guidebook of the day as somewhere "*no woman with regard to propriety or decency would care to be seen*". More "cultured" citizens enjoyed the new museum, built on the site of the George and Dragon coaching inn using funds provided by mystery man George Beaney.

James George Beaney had died in 1891 in Australia where he'd spent most of his life. He'd grown up in Canterbury in the 1830s, emigrated as a young man and then returned to study for a diploma at the Royal College of Surgeons in Edinburgh. After serving in the Crimean War he'd returned to Australia to set up a private practice. In the years that followed, several of his patients died suspiciously, leading him to be charged on three separate occasions with manslaughter and once with murder. In each case he was acquitted. During his lifetime he mysteriously acquired a fortune, leaving large sums to various educational and charitable causes in his will. This included £10,000 left to the City of Canterbury to set up "The Beaney Institute for the Education of the Working Man". Despite questions about the man's reputation, the city fathers were more than happy to accept his money, demolish an ancient inn and build a museum in its place.

As with the rest of the Western world, Canterbury was affected by an onslaught of new media around this time. Radio was becoming popular, leading to the slow decline of the locally distinctive "Kentish burr" accent. And in the 1910s Canterbury's first cinema, The Electric Theatre, opened in St. Peter's Street. Although very few could afford to buy a camera, photography was also establishing itself locally, with Richard Sinclair & Son running a studio at 101 Northgate. Local children, like children nationwide, began to follow avidly a new cartoon strip featuring Rupert Bear and a cast of his therianthropic friends, published in the *Daily Express* from November 1920. This was the creation of Mary Tourtel, born Mary Caldwell in Palace Street in 1874 and educated at Simon Langton Girls' School, then Thomas Sidney Cooper's art school in St. Peter's Street. Novelist Virginia Woolf spent the summer of 1910 at her sister's home just north of town in Blean, recovering from a mental breakdown. In 1904 she'd written to a cousin that "*there is no lovelier place in the world than Canterbury*", despite having seen Venice and Florence. In 1925 she published a short story called "Together and Apart" describing a brief, awkward exchange between a young woman and middle-aged man which centred on their both knowing Canterbury.

A memorial to members of the local "Buffs" (Royal East Kent) regiment who died in the recent

Boer War in southern Africa was erected in the Dane John Gardens in 1904. In 1920 or '21, when a site was being prepared for a new Kent County War Memorial on the western edge of the Cathedral precincts, excavations uncovered some Saxon-era foundations, believed to be those of the church of St. Mary Queningate. This would have been one of numerous churches and chapels dedicated to St. Mary in pre-Norman Canterbury. Another of these, St. Mary Bredman on the High Street, was demolished in 1900.

In 1906, Nevill Cooper, son of the famous local artist, began planting a "pinetum" (an arboretum consisting of conifer species) on some land adjacent to Alcroft Grange. This was a mock Tudor mansion which his father (who'd died in 1902) had had built on a vast tract of land atop St. Stephen's Hill, bought with the proceeds from a single painting – such had been the extent of his celebrity. This was one of two pineta which Cooper established, the other being at his father's home, Vernon Holme in Harbledown, where he also created an extensive rock garden. Locally he was known for his eccentric style of dress, his involvement with the Canterbury Old Stagers theatre group and his funding of the Theatre Royal in Guildhall Street. When he died in 1936, a newspaper obituary reported that he'd had a lifelong addiction to sleeping draughts – almost certainly a reference to laudanum, an opium tincture.

A similarly mysterious character to George Beaney got involved in local politics around 1905. Francis Bennett-Goldney inherited an unexplained fortune and quickly became an alderman, the mayor and then an MP. While mayor he also received the title of "Athlone Pursuivant of Arms in Ireland" (Ireland still being under the control of the British Crown). Two months later the Crown Jewels of Ireland vanished on his watch. When he did in a car crash in 1917 after becoming an honorary military attaché at the British Embassy in Paris, it became apparent that he had stolen several charters and other historical documents belonging to the City of Canterbury, as well as a painting belonging to the Duke of Bedford. Many questions about his life remain unanswered. Shortly before Goldney's involvement in the life of the city, the council had adopted a motto and coat-of-arms for Canterbury. The motto, *Ave Mater Angliae* ("Hail Mother of England") refers to the role of St. Martin's as the "mother church" which arguably birthed Christianity in the country and the coat-of-arms features a golden leopard and three choughs, black birds borrowed from Thomas Becket's family heraldry. At a time when the influence of religion on the life of the city was rapidly in decline, this officially fixed the dates 597 and 1170 as the two pivotal points in its history.

When war was declared in 1914 many young reservists were sent off to fight, a military band playing for them in the Dane John Gardens. During the war, Canterbury experienced German zeppelin raids but suffered only one minor casualty, someone's foot being damaged by shrapnel. But the effects of the war were certainly felt locally. Thousands of troops were quartered in the city, so pubs were asked to close an hour early. Rationing was introduced, but it wasn't that successful as there simply wasn't enough food to go around. A communal kitchen was opened at 92 Northgate where locals could bring their own bowls to be filled with cheap soup, pies and mashed potatoes. A hospital was set up in the Dane John Gardens for wounded troops.

the High Street during Canterbury's coaching heyday, based on an 1827 lithograph by Thomas Sidney Cooper, with the Paris and London Royal Mail coach and its passengers outside the Coach and Waggon Office at the George and Dragon Inn (this was demolished and replaced by the Beaney Institute in the 1890s, still the city's museum and library in 2017)

left: the Electric Theatre in St. Peter's Street, Canterbury's first cinema, sometime in the 1910s; *centre*: Valentine Sinclair's Curiosity Shoppe, later to become Richard Sinclair & Sons' photography studio, the city's first; *right*: the 1899 memorial on Martyrs' Field Road in Wincheap to the Protestant martyrs burned by order of "Bloody" Mary in the 16th century

the opening of the Canterbury-Whitstable railway line in May 1830, the carriages being pulled by *Invicta*, a locomotive built by George Stephenson in 1825 (he was a guest on honour at a celebratory feast at the King's Head Hotel that evening)

the so-called Battle of Bossenden Woods in which charismatic rabble-rouser John Nichols Thom (then calling himself "Sir William Courtenay") and seven of his followers were killed by troops a few miles from Canterbury

left: local celebrity artist Thomas Sidney Cooper (1803-1902), based on an 1867 self-portrait; *centre*: the city's coat-of-arms, created at the beginning of the 20th century, featuring a golden leopard and three choughs, the Latin motto meaning "Hail, Mother of England"; *right*: the Egyptomaniacal (or "Egyptian revival" style) synagogue built in King Street in the 1840s

left: James Simmons' unpopular brick replacement for the medieval Ridingate in 1882, the year before it was pulled down to be replaced by a simple iron bridge; *right*: St. George's Street in Edwardian times with its church and clocktower visible

7th January 1921 to 9th January 1940

The young men of Canterbury and Kent who died in the Great War were commemorated with a pair of memorials in 1921. In August, Marchioness Camden unveiled the Kent County War Memorial within the Cathedral Precincts, just inside the old Queningate, and in October, Earl Haig unveiled the city war memorial at the centre of the Buttermarket, replacing the controversial memorial to Christopher Marlowe.

Three years later, the city was visited by Scottish politician and journalist Robert Cunningham Grahame who wrote that "*the houses with their casement windows, timbered upper storeys, and overhanging eaves, still kept an air of an older world. The gateways with their battlements and low archways, through which the medieval traffic once had flowed...to the shrine of Becket, were now mere monuments. Grouped round its dominating church, the city huddled as if it sought protection against progress and modernity. Bell Harry in his beauty pointed heavenwards. It was indeed a haven.*"

In 1924 Canterbury's literary heritage received an indirect boost when the Polish novelist Joseph Conrad (often described as the greatest writer in English who wasn't a native speaker) died at his home of some years in nearby Bishopsbourne and was then buried in Canterbury City Cemetery during the annual Cricket Week celebrations. His burial was a small, private affair, very few among those throngs crowding the streets at the time even being aware of Conrad's existence. A couple of years later the cultural life of the city received a blow when the Theatre Royal in Guildhall Street finally closed (it had been heavily subsidised by the artist Thomas Sidney Cooper and later by his bohemian son Nevill). It was to be pulled down for the building of Lefevre's Department Store, the Lefevres being one of the most successful Canterbury families descended from 16th century French Protestant immigrants. By 1933, though, the locals were not lacking entertainment, as a pair of rival cinemas had opened at either end of town. Jazz was also becoming popular with the locals, dances being held several nights each week at one point. A low attendance at an orchestral concert in St. Dunstan's led the vicar, Reverend E. A. Miller, to write in his parish magazine, "*What a pity that good music does not attract people nowadays; the heathenish noises of jazz seem to be preferred.*"

In 1928, a Catholic church and college in Hackington, converted in the 1880s by French Jesuits from the former Hales family mansion and dedicated to St. Mary, began to be demolished, the Jesuits having returned to France. During this process, a number of corpses (both from the Hales family vault and from the more recent Jesuit occupation) were exhumed and reinterred in a radial pattern around a tiny, cylindrical Catholic chapel which had been converted from a dovecote around 1850. Most unusually for a consecrated building, its structure involves large numbers of animal bones. It's currently claimed to be the smallest consecrated space in the Christian world.

The first post-Saxon Archbishop of Canterbury to resign from office was Randall Davidson, in 1928. From then on it became customary for archbishops to serve for roughly ten years, including a single Lambeth Conference. These conferences, bringing together representatives of the worldwide Anglican community, had been held every ten years since 1867. Although held in Lambeth, they often included brief trips to Canterbury. Davidson was replaced by Cosmo Lang who was enthroned at the Cathedral on an "occasion of great magnificence". The celebrated English composer Ralph Vaughan Williams composed a *Te Deum* for the occasion and the Southern Railway Company laid on a special train for visitors from London. Lang fell in love with the Cathedral, writing "*Surely the view of the choir and apse at Canterbury is one of the most beautiful pictures in stone in England, perhaps in the world.*" During his tenure, *Murder in the Cathedral*, T.S. Eliot's play about the Becket affair, was performed in the Cathedral Chapterhouse as part of the 1935 Canterbury Festival, an annual cultural event established in the 1920s.

Motorised transport was becoming more common with bus services now running between Canterbury and nearby towns, the East Kent Road Car Company having been established in 1916. In 1931, improvements to the Canterbury-Whitstable bus service resulted in the cancellation of passenger rail services on the old "Crab and Winkle Line" as it had become known, although freight continued to be transported to and from Whitstable harbour by rail. A new Kent and Canterbury Hospital was built just off the Old Dover Road in 1935-37, replacing the one near the Abbey grounds, and in 1936 the gardens beside the Stour outside the Westgate, formerly meadows known as the Rosiers, were presented to the people. The picturesque gardens and riverside walks had been created by the Williamson family, owners of the tannery in St. Mildred's Parish, who'd lived there since 1886. Stephen and Catherine Williamson gifted the property to the city on the condition that it was to be used for the benefit of the residents. Tower House, their residence which incorporated a bastion tower originally part of the city wall, was converted into the Lord Mayor's office. Some years later Catherine Williamson became Canterbury's first female mayor.

Changes to the appearance of the city included the disappearance of Denne's (formerly the Abbot's) Mill, destroyed by a three day fire in 1933, having stood since 1792, the demolition of the old rectory of St. Martin's Church for road widening in 1936 and the demolition of All Saints Church (a 19th century brick replacement of the medieval original) in 1937. The small churchyard of All Saints was converted into Best Lane Garden. Poverty was still very much in evidence in some parts of the city, particularly Northgate, hundreds of whose children crowded into Westgate Hall each January for an annual "Northgate Poor Kiddies Treat".

The Second World War was to prove far more catastrophic for Canterbury than the First, but of course no one could have known that when hostilities began in 1939. Because so many young men were called up to serve in the military, King's School boys had to be brought in to pick apples from the surrounding orchards that September. The same month, lorryloads of dry earth were moved into the Cathedral to protect its interior from bomb damage and valuable stained glass was removed to be kept in safe storage. Gas masks were distributed to locals from the Corn Exchange and a warning siren was installed at the gasworks beside the castle (the council had bought the castle from the Gas, Light and Coke Company in 1928).

between the wars

9th January 1940 to 4th March 1955

German bombs began falling on Canterbury, the first casualty being recorded on 21st July 1940. Eight people died when a row of houses was bombed in September, and the next month Burgate was badly hit. Miss Carver, of Carver & Staniforth's bookshop, a well-known local character, was killed along with her cat. That August, three German bombers had been brought down in the area: a Dornier 17 in Rough Common and Messerschmitt 109's in Barham and Chilham. The next year, the conservative archbishop Cosmo Lang retreated to Canterbury from Lambeth when the bombing of London became too intense. He was to retire the next year, to be replaced by the progressive William Temple who died only two and a half years later (the last archbishop to die in office). By 1941 an anti-invasion plan was in place, including tank traps. The defensive perimeters for the city would have required a number of house demolitions but fortunately this never became necessary.

1st June 1942 saw Canterbury's worst bombing of the war, Germany's direct response to the RAF bombing of Cologne. A large part of the city around St. George's Street and the Whitefriars area was almost completely destroyed, including St. Mary Bredin church. The missionary college at St. Augustine's was also damaged, effectively ending its life, but remarkably, apart from its library being destroyed, the Cathedral escaped any major damage (volunteers stationed on its roof rushed to any incendiary bombs that landed and swept them off). Simon Langton Boys' School was badly damaged but managed to continue operating until relocating to the Nackington Road south of the city in 1959 (King's School, although not directly affected by the bombing, had relocated to Cornwall for the war's duration). The next day, the Cathedral's Dean, Hewlett Johnson (known as the "Red Dean" for his communist leanings) was seen wandering around the bombed-out parts of town, dressed immaculately and offering comfort to those affected while stray dogs ran around in a state of panic and terror. The last time anything comparable had happened to the city was the Viking siege of 1011.

Another devastating raid occurred on 31st October that year, a Saturday afternoon when the streets were full of shoppers from out of town. A bus on the Sturry Road was raked with machine gun fire. Further bombing in early 1944 damaged properties in the London Road and St. Dunstan's Street, including the church. By the end of hostilities a total of 115 people had been killed in Canterbury, with 380 injured and 800 buildings destroyed.

One small consolation for the devastated city was the possibility of archaeological discovery where previously the density of medieval buildings had meant that this was out of the question. Archaeologists soon got to work in 1945, finding the remains of a Roman house with a beautiful mosaic floor deep below Butchery Lane (the street level of the city having been creeping upward continuously for 1900 years). Further excavations in 1950 led to the discovery of the two-stage Roman theatre under the Castle Street-Watling Street crossroads. Plans to restructure the city were floated almost immediately after the war, too, including one for a relief road parallel to the high street (now a major thoroughfare for motorised transport between London and Dover). That plan, the brainchild of Dr. Charles Holden, was bitterly opposed by a newly formed Citizen's Defence Association which took on the city council and managed to gain a majority of its seats in the November 1945 local election. Had they not done so, there might still be heavy traffic thundering over a concrete flyover spanning the Greyfriars Garden and Binnewith Island. So the London-Dover traffic continued along the high street, but with the increased motorisation of society it was clear that this was not going to be sustainable. The idea which eventually came to be accepted in 1952 was for a ring road outside the city walls. Attempting to drag Canterbury out of its medieval past and into "modernity", the same plan included two new cross-city relief roads and four multi-storey carparks. Not long before, the city's 770-year-old Guildhall had been demolished, supposedly because it was beyond repair although this assessment was vigorously disputed.

The 1951 Festival of Britain, an attempt to cheer up the post-war nation, sponsored an exhibition on the history of the city in the bombed-out remains of the Whitefriars estate that June. Three months later, bulldozers turned up to remove the exhibition so that reconstruction work could begin and surprised everyone by also removing almost all that was left of the walls of the medieval friary. The next January, a Woolworths opened, the first new structure to appear in the heavily bombed St. George's Street, a symbol of modernity attracting huge queues – Woolworthsmania!

The Canterbury-Whitstable railway finally closed in 1952 but the next year it was temporarily reopened as part of a relief effort when Whitstable was hit by terrible flooding. As part of the same relief effort, Emperor Haile Selassie of Ethiopia sent the people of Whitstable a shipment of Ethiopian coffee, a tin of which is still stored in Whitstable Museum (although not on permanent display). The last rails were removed from the track in January 1954, the section through the Tyler Hill tunnel having proved particularly tricky.

The 1953 coronation of Queen Elizabeth II, officiated by Archbishop Geoffrey Fisher (the only known Freemason Archbishop of Canterbury) was commemorated locally with a "historical pageant" in the Dane John Gardens, its ancient mound surmounted by a remarkably tasteless giant crown illuminated by electric lights. Also, as for every new monarch since the 12th century, the city council's Burghmote Horn was blown to mark the event. Shortly before the coronation, the Queen of Tonga visited the Cathedral on her way to London.

In 1955 work began on a new roundabout just beyond the Ridingate. This involved the demolition of the Ridingate Inn and the closing of the cattle market which had existed in the same place for over a millennium. This also ended the annual hiring fair which had taken place there since the 1820s, replacing the one in the Precincts. Three years earlier, the *Architectural Journal* had published a scathing piece on the city officials' approach to postwar reconstruction. Tensions were starting to mount over the future of Canterbury. Opposing forces of "preservation" and "progress" locked horns over the matter of preserving the city's historical image versus improving its commercial prospects. Thankfully, the "progress" faction was forced to compromise, neither of the cross-city relief roads being built and only two of the four multi-storey carparks (one at Gravel Walk and one near St. Mildred's church).

St. Dunstan's Street as it would have looked in the 1920s, the Westgate being the only remaining medieval gate in the city wall circuit at this time

left: Rupert Bear drawn in the original style of his creator Mary Tourtel in 1921, here seen visiting Fairyland with his friend the Wise Old Goat; *right*: Rupert with his friend Bill Badger in 1948, in the style of Alfred Bestall, Tourtel's successor. It's interesting that with his appearance in the early 1920s, when traditional folklore was fading from children's lives and being replaced by modernity and technology, a simple newspaper strip involving animal-headed humanoids encoutering fairies, sprites and elves became wildly popular with the nation's children.

left: the medieval Guildhall as it would have looked prior to its controversial demolition in 1950; *right*: the original Simon Langton Boys' Grammar School in the Whitefriars area before it incurred extensive WWII bomb damage and was moved out to the Nackington Road

left: the WWI memorial being unveiled in the Buttermarket in October 1921; *right*: Archbishop William Temple outside the Archbishop's Palace after his April 1942 enthronement. His father Frederick was the first Archbishop of Canterbury (1896-1903) to live in the newly restored Palace. William Temple and his wife Frances were resident during the massive bombing raid a few months later, seen hurrying around the Precincts in tin hats and pyjamas (she even helping to hurl burning furniture out of windows to stop fires spreading).

left: total devastation caused by bombing raids in the St. George's and Whitefriars areas; *right*: bomb damage in St. George's Street, where the church was largely destroyed (the tower survived and still stands as a clocktower)

left: destruction in Burgate in July 1942; *right*: the first Marlowe Theatre, opened in 1950 in St. Margaret's Street

4th March 1955 to 29th March 1967

By 1957 the effects of rock'n'roll were being felt in the venerable cathedral city. The local press reported on 2nd October that there had been complaints about "Teddy Boys" damaging a playground on Littlebourne Road. A different order of damage was being done in the name of "progress", meanwhile, with the Fleur-de-Lys, an old coaching inn which had stood in the High Street for 600 years, being demolished in 1958. More understandably, certain areas of the city which had become slums began to be redeveloped. A slum clearance program had begun in 1938 but then been suspended during the war. In 1958 the city council accepted tender for the demolition of twenty-one houses on New Ruttington Lane and 1966 saw the mass demolition of compulsorily purchased houses in Northgate. Meanwhile, everything around the outside of the southern half of the city wall circuit was being demolished to make way for half of the new ring road. The northern half never got built as the exact route couldn't be agreed upon, although the plans weren't formally dropped until 1976, after extensive public protests.

The Canterbury Preservation Society (later to become simply the Canterbury Society) was founded in 1961 in response to the horrors of the postwar demolition and redevelopment programmes, seeing the latter as a terribly misguided interpretation of modernist architecture, completely unsympathetic to its surroundings. Over time, the city authorities began to be won over to their views, not so much because of a shared sense of taste or duty of care, but because it became apparent that the historic resonance of a place like Canterbury could now be commodified and sold. From this point on, Canterbury began a gradual slide into a sort of historical self-consciousness, a money-spinning "ye olde Canterbury" for the tourists being built on top of the day-to-day 20th century life of what a young resident of the time called Richard Sinclair would later describe as a "giant village in the Garden of England" (his namesake great-grandfather had run that early photography studio in Northgate).

After many years of planning, the construction of the "University of Kent at Canterbury" began in 1962 atop the hill overlooking the city, not far from where St. Edmund's School had relocated in the previous century. Part of the land was purchased from the Master of Eastbridge Hospital, still formally the Lord of the Manor in Blean where the 13th century High Street hospital held land from its medieval farms. UKC opened to 500 students in 1965. A single college (Eliot) was joined by three others (Rutherford, Darwin and Keynes) over the next few years. Meanwhile, in a house just a stone's throw from the site of the future Keynes College, "Tanglewood" on Giles Lane, a group of young musicians were creating something eventually to become known worldwide as the "Canterbury Sound" (or "Canterbury Scene"). The house belonged to Billie and Leslie Hopper, whose sons Hugh and Brian attended Simon Langton Grammar School. There they'd befriended Robert Ellidge whose progressive parents had taken in an Australian beatnik lodger called Daevid Allen in 1961. Daevid and Robert shared an enthusiasm for jazz and the avant-garde, which spread quickly among Robert's circle of school friends. By 1963, Daevid (guitar/poetry), Hugh (bass) and Robert (drums) were playing gigs in London as the Daevid Allen Trio.

Before long, another local long-haired young man (almost unheard of in East Kent at the time) called Kevin Ayers joined the circle of friends, being recruited to sing for a more conventional Beatles-inspired pop group called The Wilde Flowers. The aforementioned Richard Sinclair was brought in as a guitarist. Rehearsing in the Hoppers' tiny house, the band gigged around Canterbury between 1965 and 1967, most regularly at a little club called The Beehive on Dover Street. Kevin went off travelling, with Robert taking over vocals and a new recruit, Richard Coughlan, taking his place on drums. As well as playing the popular R&B of the time, this group of friends continued to take an interest in the avant-garde, experimenting with tape composition, filmmaking, etc., sometimes holding "cultural evenings" at Tanglewood when Hugh and Brian's parents were away.

On Easter Sunday 1966, Kevin and Daevid shared an intense LSD experience in Deià, Mallorca while staying at the home of Robert Graves, the poet and novelist who was also a friend of Robert's parents. This inspired them to create a new kind of music and so they returned to Canterbury to recruit Robert as drummer and, to play organ, Brian's pianist friend Mike Ratledge, another Langtonian who'd recently returned from Oxford where he'd earned a philosophy degree with distinction. This group began rehearsals at an old house in nearby Sturry that summer, the Wilde Flowers continuing with a new lineup. They eventually settled on the name "The Soft Machine" and decided to relocate to London, the centre of countercultural activity at the time. Robert's recently widowed mother, Honor Wyatt, had moved to a house in Southeast London and invited the entire group (and their girlfriends) to stay. They quickly found themselves at the centre of the psychedelic vortex known as "swinging London", playing at such legendary venues as UFO and Middle Earth. It was at this time that Robert Ellidge adopted the name Robert Wyatt.

Also making a cultural mark far beyond the bounds of the city, in 1959 the creative duo of Oliver Postgate and Peter Firmin had launched "Smallfilms", a company specialising in animated films for children's television, operating out of a converted cowshed just off the Whitstable Road in Blean. Early creations included *Ivor the Engine*, *The Pingwings*, *Noggin the Nog* and *Pogles' Wood*. The latter, featuring a family of diminutive woodland dwellers, a shapeshifting witch, the King of the Fairies and a magical storytelling plant, became wildly popular with the nation's small children between 1966 and 1968.

Around the time that a mysterious "hum" which had been troubling people across Kent (including Canterbury) was raised in Parliament – the mystery was never solved – Archbishop Geoffrey Fisher, the first to make use of air travel with trips to Jerusalem, Constantinople and Rome, met Pope John XXIII. This was 1960, the first such meeting since Henry VIII's Reformation of the English Church. The meeting was brief and low-key, with no photos allowed. The next year, Fisher was succeeded by Michael Ramsey, a surprisingly jolly character considering his reputation as a profound theologian (with "unfailing good humour and generosity of spirit"). When in Canterbury, he was known to drop in to the Bell and Crown in Palace Street occasionally for a pint and a chat with the locals. In 1966 he met the next pope, Paul VI, furthering the thawing of Anglican-Catholic relations, even being given a ring by that pontiff.

29th March 1967 to 10th November 1976

The psychedelic goings-on in London made their impression on Canterbury on 6th May 1967 when the Soft Machine returned to play the Rag Ball at the City Technical College, just off the New Dover Road, next to the art college. Joining them were The Wilde Flowers, still featuring the Hopper brothers but now completed by members of the soon-to-be-formed "Canterbury scene" band Caravan, including Richard Coughlan, Richard Sinclair and his cousin Dave Sinclair. The Soft Machine then went on to tour France where they spent a memorable summer becoming the darlings of the Parisian avant-garde, even being elected official orchestra of the absurdist Collège de 'Pataphysique. On their return via Dover in late August, Daevid Allen (an Australian passport holder) was barred from re-entering the country. This was officially for immigration reasons although he later claimed that his interrogators accused him of involvement in international LSD trafficking. Unperturbed, he remained in France and launched the extraordinary commune-based psychedelic space-jazz band Gong with whom he remained until 1975. Just over a week later, while Soft Machine were getting set up to provide music for experimental "ballet-in-a-bathtub" *Lullaby For Catatonics* at the Edinburgh Festival, artist Jean Watkins visited Canterbury with her husband Bill on a daytrip from London, little knowing that almost fifty years later she'd be illustrating its historic buildings (and events yet to come) for a book by their yet-to-be-conceived child Matthew.

Soft Machine continued as a trio (Wyatt, Ratledge, Ayers), touring the USA with Jimi Hendrix for most of 1968, during the April of which they recorded their eponymous debut album. An exhausted Kevin Ayers then sold his bass and quietly left the band. Between tours they'd been back in the Canterbury area, rehearsing at Graveney Village Hall, with Hugh Hopper (Kevin's eventual replacement) standing in on bass. Around the same time, Caravan recorded *their* self-titled debut LP having signed a record deal after an intensive period of rehearsal in the same building, behind (and, on cold nights, inside) which they'd been camping. Following in the Soft Machine's footsteps, Caravan relocated to London, but soon decided to return to a healthier and less stressful life in Canterbury where they continued to operate for a decade or so.

The revolutionary spirit sweeping university campuses across the Western world in 1968 reached UKC in the form of "FUKC" (the "Free University of Kent at Canterbury"), founded by zealous students who'd occupied the Registry Building. The next year a young guitar prodigy named Steve Hillage arrived to study at what had quickly reverted back to just "UKC". A student of the time recalled him usually walking around in a wizard's cloak looking "very stoned". Hillage lived in a shared house at 5 St. Radigund's Street, a centre for psychedelic counterculture in the city, jammed with members of Caravan, was then introduced to Kevin Ayers and joined his band, leading him to Gong's commune in France where he became part of their classic 1971-74 lineup. Also living at the St. Radigund's house were members of "prog-folk" band Spirogyra. They were UKC students, but signed by a record company, and they spread awareness of this little-known Saxon female saint's name by releasing an album called *St. Radigunds* in 1971. Singer Barbara Gaskin was introduced by Hillage to an old London bandmate, Dave Stewart, who she eventually married after they worked together in the band

Hatfield and the North with Richard Sinclair following his 1972 departure from Caravan. The University itself provided a venue for touring bands, with Led Zeppelin playing the Sports Hall in 1971 and, the same year, Soft Machine returning in their classic quartet lineup (locals Wyatt, Hopper and Ratledge plus Elton Dean on saxes) to play Darwin College dining hall.

Wyatt left the Soft Machine not long after, starting a new group called Matching Mole with Dave Sinclair from Caravan plus Phil Miller from Hatfield and the North, then went on to have an extraordinary solo career. Kevin Ayers recorded a series of weird and wonderful solo albums from '69 to '78, Mike Ratledge and Hugh Hopper continued with the Soft Machine, and a cluster of musicians around them played in endless permutations, creating an astonishing amount of highly original music, now collectively known to aficionados as "Canterbury" music, even though a lot of it originated in London and elsewhere. In the city itself, not much of this music was happening by the mid-70s, with the exception of The Polite Force, a very "Canterbury" sounding jazz-fusion band involving Dave Sinclair and Graham Flight, the latter having briefly been Robert Wyatt's replacement as singer in the Wilde Flowers.

Over 400 years after the Greyfriars estate in the city centre was dissolved, up on the hill, just along Giles Lane from the UKC campus and the Hopper family house, friars following the Rule of St. Francis of Assisi re-established themselves at the new Franciscan International Study Centre in 1974. The same year, the 1820s Canterbury-Whitstable railway tunnel running though the hill collapsed, causing part of UKC's Cornwallis building to subside into it. Around the same time, the city lost its status as England's smallest county borough, effectively breaking a promise made by Edward IV just over 500 years earlier. This was the result of 1974 legislation reforming the structure of local government in the UK. In order to return it to "city" status, a supplementary charter was issued by the Queen on the 28th of May. Meanwhile, up in the old cowshed in Blean, Postgate and Firmin's "Smallfilms" operation was experiencing its heyday, producing the extremely well-loved children's TV series *Bagpuss* and *The Clangers*.

Free festival culture was on the rise in England in the years following the first Stonehenge Free Festival in the summer of 1974. In the summer of '76, the Tangmere Free Festival, planned for a site in West Sussex, was legally blocked from happening and the organisers, seeking a new site, somehow settled on the Broad Oak Valley a few miles northeast of Canterbury. Revellers arrived to find police roadblocks, leading to a few arrests for drug possession. The festival was moved at the last minute to coastal land near Seasalter. Most of the planned acts failed to appear but the crowd was entertained for three nights by former Gong synthesiser wizard Tim Blake with his pioneering "Crystal Machine" laser setup, at one point asking the audience to meditate on the generator so it wouldn't run out of petrol. Meanwhile, Hells Angels bikers took care of the dope dealing.

Although spared from the main London-Dover traffic since the construction of the half-ring-road in the 60s, the high street wasn't pedestrianised until the mid-80s. The first stage of this process was accomplished in 1975 when the St. George's Street portion became traffic-free.

left: Daevid Allen outside Wellington House, Robert Wyatt's family home in the village of Lydden, 1960 or '61; *right*: The Wilde Flowers' original lineup: Brian Hopper, Richard Sinclair, Robert Ellidge (later Wyatt), Hugh Hopper, Kevin Ayers

left: the Soft Machine posing in a London park in 1966 or '67: Kevin Ayers, Robert Wyatt, Mike Ratledge, Daevid Allen; *right*: the classic Caravan quartet around 1968: Pye Hastings, Richard Sinclair, Richard Coughlan, Dave Sinclair

left: Tanglewood on Giles Lane, home of the Hopper brothers and the Wilde Flowers' rehearsal space; *right*: 5. St. Radigund's Street, the unusually shaped house on the corner of Duck Lane, home to future Gong guitarist Steve Hillage during his brief spell as a UKC student, as well as the cult prog-folk band Spirogyra

left: Steve Hillage and Kevin Ayers (wearing a tea cosy?) contemplating a Jim Reeves LP; *right*: a police roadblock set up to prevent hippy revellers from holding a free festival in the Broad Oak Valley in the summer of '76

left: the first batch of undergraduate students arriving at Eliot College in 1965, the year the University of Kent at Canterbury opened; *right*: damage caused to UKC's Cornwallis Building in 1974 due to parts of the structure subsiding into the old Canterbury-Whitstable railway tunnel

left: Bagpuss and friends: Gabriel the banjo-playing toad, Professor Yaffle the woodpecker (a character modelled on the philosopher Bertrand Russell) and the adorable mice; *right*: Smallfilms creators Oliver Postgate and Peter Firmin working on *The Clangers* in their studio in Blean (a converted cowshed)

10th November 1976 to 11th July 1984

A clearer picture of Canterbury's past was emerging with increasingly detailed archaeology, the Canterbury Archaeological Trust having been founded in 1975. In 1977 a remarkable find was made: the skeletons of two Roman soldiers in a site near the Norman castle keep, twisted in such a way as to suggest a hasty burial, bodies having been thrown in a pit with their swords thrown in on top of them. Although we'll never know the circumstances, this was clearly not a ceremonial burial, so a likely scenario involves a murder of the two soldiers and a swift disposal of their bodies. This was the first (possible) indication of any resistance to Roman rule from the native population. In 1982 an excavation on the London Road revealed the re-occupation around 600CE of a Roman cremation cemetery by Jutish or Saxon settlers after the period of desolation. Coins, glass cups and reused Roman vessels were found, as well as a beautiful gold pendant displaying a high degree of craftsmanship and integrating a cross into a more traditional pre-Christian design, suggesting the work of a local jewellery maker for a wealthy Christian customer (pictured on p. 51).

The communal house in St. Radigund's Street, a centre of the city's hippy subculture since the late 60s days of Steve Hillage and Spirogyra, was sold in 1977, ending that particular era, a few punks now starting to be seen around town. That January, a baby was born in Canterbury and named Orlando Bloom, later to attend St. Edmund's and The King's School, and one day destined to portray the world's most famous elf to an audience of tens of millions. He was named after the composer Orlando Gibbons who'd died in Canterbury in 1625 and was buried in the Cathedral. The King's School, arguably the oldest in the world (if its claim to be directly descended from the original Abbey school is taken seriously), broke with centuries of tradition in the 1970s when girls were first admitted to the sixth form, its first step towards becoming fully coeducational.

In 1978 Canterbury City Council finally got a new Guildhall, almost three decades after the 12th century original was demolished: Holy Cross Church beside the Westgate. This had become redundant five years earlier, having served as an interdenominational chaplaincy for the University since 1966. It was formally opened by Prince Charles who was made a freeman of Canterbury as part of the ceremony. That summer, the eleventh Lambeth Conference was held on the UKC campus (as have all subsequent conferences). The conferences had previously been held in Lambeth, once each decade since the late 19th century. 440 bishops attended from around the Anglican world, topics under discussion including capital punishment and the rights of individual churches to decide on the matter of female priests.

In 1980 the nation was televisually transported to Canterbury via Mike Leigh's BBC TV film *Grown-Ups*, filmed on location in the city. This was to be the first television film to be screened at the annual London Film Festival. A couple of years later, Canterbury was back on the nation's screens, this time in a much more prominent and non-fictional context. Pope John Paul II became the first pope ever to set foot on British soil (other than Adrian IV, the only English pope, in the 12th century). He paid an official visit to Canterbury in May 1982 where he joined Archbishop Robert Runcie in silent prayer at the site of Becket's martyrdom in the Cathedral. Huge crowds were predicted, local shopkeepers expecting to make major profits that day, but these predictions only led the (arguably contrary) people of the city to stay at home and watch the events on television, if at all.

18th January (Epiphany) 1981 saw a daring theft take place in the Cathedral crypt. An ivory statue of the Virgin Mary was stolen from the Chapel of Our Lady Undercroft. It was of 17th century Portuguese workmanship, having been gifted to the Cathedral in 1948 by Leslie de Saram, a wealthy Sri Lankan lawyer and politician. No one was ever caught and the statue (then valued at £2500) was never recovered. The delicate base from which it was broken has since remained in deep storage in the Cathedral. The chapel, dating back to at least the 14th century, possibly the 9th, lies almost directly below the Cathedral's central altar and during the pilgrimage heyday had been richly decorated with twinkly silver foil. The statue niche was to remain empty until the summer of '82 when a newly commissioned sculpture of the Madonna and child was installed. This was a 3 foot 6 inch bronze piece created by Mother Concordia, a Scottish nun formerly known as Carolyn Scott, at that time Prioress of St. Mildred's Priory in Minster-in-Thanet. St. Mildred (or Mildryth), incidentally, was King Ethelbert's great-granddaughter. On the same Sunday in January '81, a 17th century Dutch wooden Madonna and child valued at £2000 was stolen from the church of St. Mary the Virgin in the nearby village of Elham, an act quite possibly carried out by the same person(s).

The national coalminers' strike in 1984 affected the nearby villages of Aylesham, Adisham and Snowdown where small coalfields had been discovered early in the century, bringing northerners with mining experience to the area. Left-leaning residents of the largely conservative-voting Canterbury joined the strikers in pickets and protests, the punk-inspired socialist singer-songwriter Billy Bragg being spotted busking in the high street around this time.

The Odeon cinema on The Friars (known from 1933-1946 as the Friars Cinema) closed on 17th October 1981 after showing the William Hurt thriller *Eyewitness*. The Marlowe Theatre on St. Margaret's Street (opened in 1950) had become insolvent, so the city council took the business over, tore the building down to build a shopping arcade and bought the defunct cinema to house a future Marlowe Theatre. The refurbishments lasted a couple of years and the theatre opened on 8th July 1984. In its last few years as the Odeon, the cinema had doubled as a particularly lively music venue, hosting concerts from many of the big names of the punk and new wave eras: The Ramones, The Clash, The Buzzcocks, The Stranglers, Joy Division, The Undertones, Siouxsie and the Banshees, Blondie, The Human League, The Pretenders, Madness, Adam and the Ants, Elvis Costello and Ian Dury, the latter having been a lecturer at the local art college a few years earlier.

Like the theatre it replaced, the new shopping arcade was also graced with the name of Canterbury's greatest literary son. It's easy to imagine Christopher Marlowe being pleased by the idea of a Marlowe Theatre operating in his city of birth centuries after his passing, less so to imagine what he would have made of the Marlowe Arcade.

The Clash

11th July 1984 to 21st August 1990

In September 1984, an unusually structured Sainsbury's supermarket opened on Kingsmead Road, its roof suspended from poles by steel cables. Despite its innovative design winning awards, not all local residents were impressed. The architects argued that the steel uprights "echoed the towers of the Cathedral", whereas the unconvinced locals joked that it would look alright once the roof was put on. Later that year, South African bishop Desmond Tutu stopped off in the city to preach in the Cathedral (and shop in Sainsbury's?) before flying on to Oslo to receive a Nobel Peace Prize for his anti-Apartheid campaigning.

Heralding major changes to come, in February 1986 Prime Minister Margaret Thatcher and French President François Mitterrand met in the Cathedral to sign the "Treaty of Canterbury", finalising the details of the Anglo-French cooperation necessary to open the Channel Tunnel between Dover and Calais, effectively changing the territorial borders of both countries. The next autumn a different kind of change came very suddenly when a hurricane which meteorologists had somehow failed to predict tore through East Kent, flattening countless trees, including almost two-thirds of the old conifer specimens planted up at Littlehall Pinetum by Victorian celebrity artist Thomas Sidney Cooper's son Nevill earlier in the century.

More African representatives of the Anglican church were seen in Canterbury in 1988 when the Lambeth Conference was again held at UKC, the University having realised that it could profit each summer by hiring out its student rooms to conference attendees. That October, one such room became my first independent home when I arrived in Canterbury to study mathematics. This was on the ground floor of Rutherford College with a view down the hill of the Cathedral towering over the city. Less interested than most in typical student pursuits, I quickly came to love the city, seeking out music, art and culture, as well as exploring the web of public footpaths through the woods around the nearby villages of Tyler Hill, Blean and Broad Oak.

The next summer, working in the SPCK (Anglican) bookshop on the recently pedestrianised high street and renting a room in St. Peter's Grove, I had my first experience of British free festival culture, soon to be almost entirely stamped out by the Conservative government in response to the post-'88 emergence of the rave scene. A small festival was held on the land behind Brookside, an old farmhouse in the Broad Oak Valley occupied for some years by bohemian students and ex-students, far enough from the nearest neighbours to be an ideal party location. This went on over a weekend in August and happened to coincide with the most spectacular meteor showers I've ever seen. I can vividly recall a succession of local (and other) bands playing on a ramshackle stage built out of palettes, scaffold poles and tarps, people wrapped in blankets sitting around fires, drinking, smoking, tripping, ooh-ing and aah-ing at the frequent meteor streaks after dark, and an absurdist "superhero workshop" put on by the cheerful five-person anarcho-squat collective calling itself the Tankerton Dance Trio.

I had to work in the bookshop on the Saturday, so got up early enough to walk back into town and smarten myself up. Standing at the counter suffering from very little sleep (and sudden cultural dislocation) I noticed a self-important clergyman with some kind of servant carrying his bag, idly flicking through books and adding them to the pile his man was carrying ("*Ah, yes, I wrote the foreword in this one...*"). Not being too sympathetic to his manner, when his books were deposited on the counter and he instructed me to put them on his account, I abruptly blurted out something like "*Well, you'll have to tell me who you are!*" Instantly, my co-worker Phillip (a theology graduate) rushed up, brushed me aside and addressed him as "Your Grace". It was Robert Runcie, the 102nd Archbishop of Canterbury. I headed back through the woods to Brookside that evening for more peaceful anarchy, shooting stars and rock'n'roll, excitedly sharing my story with newfound friends.

I was also becoming aware of a kind of mutant hippy/punk/traveller protest culture which was starting to crystallise around the resistance to the Thatcher government's unpopular new Poll Tax. In March 1990, around the time I was starting to get deep into revision for my final exams, I saw a merry band of such types assembling for a group photo in the high street shortly before marching off to London for a national protest (which became a major riot and spelled the end of the Poll Tax). They were styling themselves as a kind of recreation of Wat Tyler's Peasants Revolt of 1381 and some of the Tankerton squat crew who I'd been getting to know were among them. I felt torn between mathematics and anarcho-resistance but my prudent side won out and I didn't have the nerve to join the march, instead immersing myself ever more deeply in complex analysis and topology theorems. A few weeks later, just after my exams, I got to see some last vestiges of the Canterbury Scene of yore when the reformed classic Caravan lineup and Hugh Hopper's band played at the student-organised "Canterbury Summer Festival" at Merton Farm. At the time I was living at Woodlands, a house of Giles Lane, unaware that I was just a minute's walk from Tanglewood, the Hopper family home which had been the first rehearsal space for the key Canterbury Scene musicians.

In 1989 parts of Canterbury (the Cathedral, Abbey grounds and St. Martin's Church) were granted "World Heritage Status". Although this was primarily due to the role of the city as the birthplace of the English church in 597, it's worth considering whether this status would have been granted were it not for the murder of Becket and subsequent splendour afforded by 350 years of pilgrimage to his shrine. The bones of Becket were the pivot on which these events turned. Pilgrims didn't just come to visit Canterbury, they came to be close to his bones, widely believed to possess magical powers. And the fate of these bones is still a matter of controversy. In the early hours of 14th August 1990, ex-French Foreign Legionnaires Peregrine Prescott and Risto Pronk were arrested in the Cathedral Precincts on suspicion of attempted burglary. When taken to court three days later they pleaded guilty and were given a twelve month conditional discharge after explaining that they weren't common burglars looking to steal anything but simply trying to prove a hypothesis that Becket's remains were still existent and hidden inside the awkwardly positioned "temporary" tomb of Cardinal Odet de Coligny in the Trinity Chapel. Cardinal de Coligny was a Huguenot who'd supposedly died while visiting Canterbury in 1571 but the pair of young adventurers told the court that they believed he'd faked his death in order to safely escape the country. One of the magistrates described this as "the most remarkable explanation that we have ever heard in this court".

21st August 1990 to 5th July 1995

In early September 1990 a mysterious letter was published in the local press, attributed to "Thomas Chough" and concerning the recent arrest of the two ex-Foreign Legionnaires who claimed they were seeking to reveal the resting place of Becket's bones. Clearly the writer was using a pseudonym – the chough is a black bird which appears on Becket's family heraldry (later adopted for Canterbury's coat-of-arms). The letter argued that the men's theory wasn't plausible since its author knew the true location of the relics, buried secretly in another part of the Cathedral. The mystery writer claimed that he was one of a handful of people in each generation who were passed on this knowledge and that they gathered there each 29th December and 12th July (anniversaries of Becket's murder and translation) to pray for the reconversion of the English church to Roman Catholicism.

Having been approached about continuing on to a PhD by Roy Chisholm, one of my favourite maths lecturers, I decided to stay on at UKC. For a few weeks towards the end of the summer of 1990 I shared a house in Roper Road, not realising that Richard Sinclair from Caravan lived a short distance away across the street. The classic lineup of the band had been rehearsing there earlier in the year for a TV appearance which their Canterbury Summer Festival slot had been a warmup for. That October, while I was getting to grips with my new research topic, a bronze "Welcoming Christ" sculpture was installed above Christchurch Gate, replacing the statue of Jesus which had been destroyed by cannon shot when Oliver Cromwell's men occupied the Cathedral in the 17th century. Created by the German sculptor Klaus Ringwald, it was criticised for being disproportionately large and unwelcomingly miserable.

In December 1990 a group of dedicated volunteers set up "Canterbury Open Christmas" to provide food and shelter for the local homeless from Christmas Eve until the morning of the 27th. This was held at the Prince of Wales Youth Club on Military Road and catered to twenty-eight homeless people. As the winter weather turned particularly harsh in early '91, some of those involved decided that an emergency winter shelter was needed. This was opened soon after in the church hall of the Parish of St. Martin's and St. Paul's (Longport) and led to the idea of a permanent day centre and night shelter for the homeless. The day centre was opened in the Prince of Wales Youth Club on November 21st where it was to remain for a few years, despite quite a lot of grumbling from neighbourhood residents, largely about the clients' dogs.

In early 1992 the Cathedral nave was refloored, allowing archaeologists to get a much better idea of the size, shape and exact location of the original Saxon cathedral and even some clues about the original church St. Augustine had had built there. Later that year, Longmarket (a side street off the high street) was redeveloped, its inappropriately designed post-WWII structures having been torn down to make way for new buildings which attempted to echo medieval pilgrims inns and the gables introduced by Flemish immigrants. Although cleary an improvement on what it replaced, the scheme was criticised for its overblown scale and for blocking views of the Cathedral. In 1993 the Marlowe memorial moved once more, from the Dane John Gardens to, appropriately, a spot outside the Marlowe Theatre in The Friars.

After years of work and anticipation, the Channel Tunnel finally opened on 6th May 1994. The face of Canterbury, particularly its high street, changed noticeably as a result. Numerous coach parties of European tourists were arriving every day, with shops adapting to cater to them with signs in French, "We accept Francs" signs and previously unpretentious chip shops now offering "Traditional English Fish and Chips" to visitors looking to experience English culture in an afternoon. The coach park at Longport was unable to cope with the influx and so in April 1995 the mayor opened a new £320,000 coach park at Kingsmead while being shouted down by an angry coach driver about parking prices and the distance people would have to walk into town. Intended as a temporary measure, it was to remain there for almost fifteen years. A significant proportion of the tourist hordes consisted of French teenagers on school trips, resulting in the appearance of large numbers of street traders selling cheap souvenirs, jewellery, etc. from wheeled carts. These began to clog up the high street to the general dismay of the locals (some of whom also complained bitterly about unsupervised youth running amok) and established businesses complained about loss of revenue. As they held legitimate street trading licenses, though, the city council were unable to find a legal way to reduce their presence.

In the spring of 1995 the Cathedral controversially announced that it would begin charging a £2 entrance fee (although locals and those coming to pray were exempt). They claimed that the 2.25 million visitors each year donated an average of just 12p each. This led to a debate within the Church and within the city, but when the day came in June for the new payment regime to take effect, an unexpected power cut – caused by flooding, legally describable as an "Act of God" – meant that the first visitor had to be admitted for free.

Visible signs of the global rise of hiphop culture appeared in the form of a new wave of graffiti on the city's walls, resulting in a "Pride in Canterbury" citizens group being launched in May 1995, determined to combat what they saw not as street art but as simple vandalism. When the 50th anniversary of VE day had come and gone a couple of weeks earlier, some elderly and conservative residents expressed their disgust at the lack of flags and patriotism evident in the city ("Where now is the heart of our city?", "Canterbury should hang its head in shame", etc.) The people of Canterbury seemingly didn't want to know. Likewise, the annual carnival, an outdated feature of city life which had survived since the 1940s, was beginning to suffer from severe popular apathy.

Having completed my PhD in the summer of 1994 and then spent a year doing research at a university in Belgium, I returned in the summer of 1995 to help out with a protest camp which had sprung up on the outskirts of Whitstable to oppose the building of a multi-lane bypass of the Thanet Way/A299. This road was eventually built and destroyed a swathe of beautiful countryside I'd known well and loved. I passed through East Kent periodically for the next few years, mostly visiting friends in Whitstable. Often I'd find myself at free outdoor parties in the area put on by tVC, a Whitstable-based sound system which had come into being a few years earlier as a purveyor of eclectic party sounds, but having by this time become completely house-and-techno-oriented.

rave culture

5th July 1995 to 22nd May 1999

The "crusty" subculture emerging across Britain started to make itself noticeable in the streets of Canterbury. Protest camps had sprung up both to the north of the city (opposing the bypass near Whitstable) and to its south (opposing a development in Lyminge Forest) with protestors and hangers-on becoming noticeable in town at times, along with a new wave of beggars of varying degrees of homelessness. In 1995 a travellers' site out at Pennypot Woods near Chartham was dispersed by the council who ended up paying its residents hundreds of pounds to repair and fuel their vehicles on the condition that they leave.

In the autumn of 1995 it became clear that the homeless day centre couldn't stay on at the Prince of Wales Youth Club. It was moved temporarily to Kingsmead cricket pavilion and then on to a building near Canterbury East station in late '98 with an official opening on 31st March 1999. Free breakfasts of toast and porridge were provided to thirty or forty clients daily, along with washing facilities, addiction counselling and opportunities to learn to read and to get help with physical and mental health. The building had formerly housed a short-lived hippy-ish indoor market called The Labyrinth in which my old UKC friend Nick Dent had run his curiously named Dead Dog Cafe in 1994. By the end of '99 the company name had been changed from "Canterbury Open Christmas" to the more relevant "Canterbury Open Centre".

A general gloom about the declining ambience of the city (excessive tourism, tacky street traders, beggars, drunks, litter, graffiti, "yob culture", etc.) gave rise to some enthusiasm for a new system of CCTV cameras installed around the city centre in late '97. Also, after some failed attempts, the council managed to push through a by-law to crack down on the consumption of alcohol in the city centre, something which was becoming ever more noticeable. Initially, shops and off-licences had been asked to stop selling Special Brew and Tennent's Super, but to no avail. A survey at the time suggested that locals supported the street-drinking ban, but the UKC Law School attacked it as draconian and unworkable.

Despite all this, with the huge popularity of the recently introduced National Lottery, funds became available to improve the three "World Heritage" sites (the Cathedral, the Abbey and St. Martin's churchyard). £1 million was also spent on the Dane John Gardens: the mound was restored and a replica of the long-vanished bandstand was installed. The latter was only possible because the plans for the original bandstand had been discovered in a foundry in the north of England. Also added to the gardens was a rather abstract fountain carved from an eleven-tonne block of Portland stone. The Abbey's new visitors' centre and museum was opened on 25th May 1997 by Archbishop Carey to tie in with the 1400th anniversary of St. Augustine's arrival. A group of primary school pupils performed two pieces of Abbey-themed music, including a rap written by eleven-year-old Zoe Kay Bradley. The Archbishop told the kids that he was sure that "*St. Augustine would have been delighted with that rap*".

From the early Internet emerged the World Wide Web, its impact soon being felt in Canterbury as elsewhere. The local press announced plans for the city's first cybercafe in

Rose Lane: the "Hard Drive Cafe" was meant to open in November 1996, offering "the chance for customers to plug into the Internet, ready to surf the superhighway of cyber space". Chaucer Tech school on Spring Lane was already offering the public the chance to "surf the Net" each Wednesday evening from 6:30-8:30.

I'd first encountered the Internet in 1993 while working on my PhD at UKC, email and academic Usenet newsgroups quickly becoming invaluable research tools. I'd since been exploring the early Web while in Belgium and was fascinated by its potential. In the summer of 1996 I returned to the area and spent a couple of months camped in some woods below an orchard in Broad Oak while using the computer network at the UKC campus to set up an online parapsychology research website. This was with the support of a maverick theology lecturer in the Religious Studies department, in which I was made an "honorary research fellow". I can recall attending gigs at The Penny Theatre during this period, including a weekly folk night.

In March '98 the famously risqué white reggae artist known as Judge Dread played his last ever gig at the Penny. Real name Alex Hughes, from Snodland near Maidstone, he was perhaps best known for having his innuendo-laden songs banned by the BBC. He collapsed and died while walking off stage, aged 52, his last words being "*Let's hear it for the band!*" Regular live music didn't last much longer at the Penny. The owners claimed that they'd been trying to support local bands but couldn't make it pay so ended up transforming the place into a much more lucrative student-oriented pub the next summer.

During Archbishop Carey's 1998 Easter sermon in the Cathedral, activist Peter Tatchell famously interrupted to protest the primate's opposition to gay marriage. Although this made national news and sparked extensive debate, when he was tried under the Ecclesiastical Courts Jurisdiction Act 1860, Tatchell was fined a mere £18.60. When Carey presided over the thirteenth Lambeth Conference, held on the UKC campus that summer, the main issue was homosexuality, the bishops passing a resolution condemning it as "incompatible with scripture". This was largely driven by the numerous bishops from African countries, with Nigerian Bishop Chukwuma even attempting to exorcise the "homosexual demons" from the Reverend Richard Kirker, a leader of the Lesbian and Gay Christian Movement who'd been distributing leaflets. The next spring, the radical vicar Jonathan Blake was arrested for attempting to nail a 95-point thesis to the Cathedral gate protesting Church corruption.

When the demolition of the much loved (and council-owned) Westgate Hall was proposed, protests sprung up to save it for the community. Not long after, the Kent and Canterbury Hospital also came under threat – not of outright closure, but of quite severe cutbacks – and large-scale, sustained protests were organised. Meanwhile, the council were preparing to construct a vast new shopping development in the Whitefriars area. Chapman Taylor architects described the contemporary buildings they'd designed as "undeniably Canterbury". These were supposed to include a library, civic amenities and a new church (none of which ever appeared). Three other firms of architects had submitted proposals but Labour councillors attacked the consultation scheme, describing it as a "sham".

Internet

22nd May 1999 to 25th June 2002

The demolition of ugly, unloved post-WWII buildings in the St. George's area began in late '99 to clear the way for the new Whitefriars shopping development. By the next summer the area was looking rather like it had after the 1942 Blitz. Archaeologists moved in eagerly and launched the "Big Dig", the biggest urban archaeological excavation ever undertaken in the UK. Nothing particularly striking was unearthed but the project was seen as a success for both archaeology and tourism with many thousands flocking to the temporary Big Dig visitors' centre and exhibition. As part of this, Channel 4's *Time Team* programme came to the city to broadcast live over the August Bank Holiday weekend.

Canterbury's traffic problems had emerged almost as soon as motorised transport first appeared in the ancient city. They reached a crisis point in the autumn of 2000 when daily gridlock was occurring on the outskirts and on the ring road, particularly near the Wincheap Roundabout. Everyone seemed to have a different idea as to how to solve this so very little progress was made on the issue. Likewise, disputes over the contentious issue of coach parking ground on. Longport was temporarily reinstated as a second coach parking area but this proved disastrous and unpopular. At the opposite end of the transport spectrum, a new cycle route had opened in October 1999 between Canterbury and Whitstable, part of the national Sustrans network, largely following the route of the old railway line. Hundreds of enthusiastic cyclists cycled in either direction on the day of its opening.

Regular deaths caused by heroin led the city's coroner and the manager of Canterbury Open Centre to speak out about the drug "flooding the streets", how easily available it was and the possibility of contaminated batches being in circulation. After a syringe was found under a pew in St. Martin's, along with evidence of a small fire having been lit inside the church, the parish authorities decided that they could no longer keep its doors open to visitors. After a few months, though, a team of voluntary attendants was organised to keep it open for a few days each week.

In early 2001 residents in the area of the Marlowe Theatre won a battle against the council about the siting of an Orange mobile phone mast on top of the building. Almost immediately after this, residents in Roper Road began a struggle with Vodaphone over a mast in their street. Another source of complaints from nearby residents – for centuries in this case – was the tannery at St. Mildred's which had been operating continuously for over 220 years (and was almost certainly on the site of earlier tanneries). It was closed in early 2001, partly due to new environmental regulations aimed at trying to reduce the amount of foul-smelling gasses released by such facilities. Although dozens of jobs were to be lost, very few residents lamented the loss of the tannery itself. A debate then began as to what should be done with the site. That autumn the much despised Gravel Walk multi-storey carpark, built in a "brutalist" style in 1969 and regarded as among the ugliest structures in Britain, was finally torn down, Lord Mayor Fred Whitemore enthusiastically posing for a photo with the demolition machinery.

In the summer of 2001 some city traders began to protest the city "dying in front of our eyes".

They claimed that trade had dropped by 50% due to coach parking issues, car parking fees, street pedlars and the lack of promotion of Canterbury. Meanwhile, some members of the public complained that the visiting French teenage hordes were turning the city into a medieval theme park. But it wasn't all medievalism: by the end of the millennium, both the council and the Cathedral had launched websites, and before his retirement in 2002, George Carey became the first Archbishop of Canterbury to have his own site. In March 2000 the city's first female parish priest was appointed (Noelle Hall, of St. Martin's and St. Paul's), several months after the Cathedral appointed Reverend Susan Hope as its first female "Six Preacher", these being an institution created immediately after the Reformation in the late 1530s by Archbishop Cranmer.

After the infamous attacks on New York and Washington in September 2001, the local authorities, as elsewhere in the West, went on general alert. A month later a man of "Arab-looking appearance" was seen scattering some kind of powder in the Cathedral crypt and ran away when confronted. Fears of an anthrax attack led the Cathedral to be evacuated and a biohazard team brought in but it turned out that he'd been scattering cremation ashes. Capitalising on the aftermath of 9/11, British National Party activists were seen outside the Cathedral gate distributing anti-Islamic pamphlets. A year before it had been a Hollywood studio rather than an extreme-right party flirting with Canterbury Cathedral: Warner Brothers were hoping to film scenes for *Harry Potter and the Philosopher's Stone* there but were turned down, the Cathedral authorities expressing concerns about the film's "pagan themes".

The fate of the K&C Hospital continued to hang in the balance. Plans to downgrade it and close its accident and emergency department resulted in ongoing demonstrations of public support for the hospital. It was clearly suffering from a funding crisis, with some patients left on trolleys in corridors for days due to a lack of beds. According to two surgeons who spoke out in 2002 the crisis could be traced back to the closure of the small satellite Nunnery Fields Hospital the year before. While visiting the area from my then home in Exeter in late February 2002, I went to Whitstable to visit Nick, an old Marxist friend from my UKC days (and former proprietor of the Dead Dog Cafe) and ended up in the middle of an affiliated demonstration. Noisy and passionate, the strength of feeling was palpable.

The council decided to put Canterbury forward as a candidate for the 2008 European City of Culture but despite much fanfare in the local press it failed to even make the shortlist. The campaign slogan was the rather vague "Canterbury Can" and part of the bid involved the claim that it was the "most European part of Britain". Cynics argued that the bid was a complete waste of money as the city clearly never stood a chance.

In late March 2002 Queen Elizabeth visited the Cathedral to take part in the Maundy Thursday service, carrying out her traditional role of dispensing "Maundy money" (specially minted coins for the occasion) to seventy-five semi-randomly selected members of the community, something she'd last done in 1965. These days, no one's feet get washed by the monarch, as was once part of the tradition. Less than forty-eight hours later she, together with royalists across the nation, was mourning the death of her mother.

25th June 2002 to 11th December 2004

A new archbishop was enthroned in February 2003. Rowan Williams, former Archbishop of Wales, had already spoken out strongly against the US/UK-led invasion of Iraq and had a reputation as a formidable theologian and moral philosopher. And, unlike his predecessors, many people found him to have a warmth, almost a *cuddliness* – to the extent that the satirical Christian online magazine *Ship of Fools* celebrated his enthronement with a limited edition Rowan Williams teddy bear, complete with his famously bushy eyebrows. The fact that he wrote the foreword to an Incredible String Band songbook, a thoughtful and glowing endorsement of the cult psychedelic folk band, suggests that via his 1960s youth he'd brought a new kind of consciousness to Augustine's Throne. Just a month earlier, a small colony of Franciscan friars was established within the city walls, the first since Henry VIII dispersed the friaries in the 1530s. Brothers Austin, Bernard and Colin were Anglican Franciscans, so they weren't affiliated with Rome like their Canterbury predecessors, but they moved into the old Franciscan grounds on the Stour which includes Binnewith Island and the only remaining building of the medieval Greyfriars estate. Their responsibilities were to the City Centre Parish (an amalgamation of several earlier parishes) and to the Eastbridge Hospital of St. Thomas, an almshouse since 1150, still housing nine residents.

Meanwhile, over at the former site of another friary, that of the Whitefriars in St. George's Parish, the archaeologists had packed up and gone by August 2002 and work was beginning on the huge new shopping development. The associated bubble of civic optimism was burst that November, though, when the promised multi-million pound library and learning centre promised by Kent County Council was dropped from the plans. This was meant to have been the key "community element" of the development. The next February, the council attempted to make amends by promising to invest in the neglected old library at the 19th century Beaney Institute. On the other side of the city, independent traders were becoming concerned about the impact Whitefriars would have on their businesses as it was inevitably going to pull the commercial centre of gravity south. Getting proactive, they rebranded Palace Street, The Borough and Northgate as "The King's Mile", launching the concept with a street festival in June 2004. An experimental pedestrianisation scheme for Palace Street then began, popular with the traders but less so with drivers. The idea being pushed was that of a "continental" street cafe culture with a lot of quirky little independent shops to complement the large chains over at Whitefriars.

After six years of uncertainty, the battle to save the Kent and Canterbury Hospital from being severely downgraded achieved a partial victory in July 2003. A 24-hour local emergency centre was to remain and only 59 of 455 beds would be lost. Another public-health-related battle began, this time against a proposed SWERF (solid waste and energy recycling facility) on the Broad Oak Road. Bretts, the company behind it, stressed that it wasn't going to be an incineration plant, but opposition remained fierce and the proposal was withdrawn in late 2003. The growing popularity of mobile phones meant that ever more microwave antennas were being installed around the area at this time. The council seriously considered an offer to have a T-Mobile mast installed in the Westgate but residents organised to stop this, fearing unknown health risks, a similar protest occurring in Blean against a proposed Orange network mast in September 2004.

Drug use was changing in Canterbury with people reporting that cocaine was now showing up everywhere in the social life of the city, previously a relative rarity. More ominously still, it was also showing up in the form of crack. The Scrine Foundation, linked to the Canterbury Open Centre for the homeless, warned that Canterbury was being targeted by ruthless gangs of crack dealers from London and Dover who were attempting to create a demand. The general feeling that "drugs" were the problem led to a curious reaction from the local police: visiting various shops selling perfectly legal paraphernalia – almost entirely cannabis-related – seemingly just to intimidate them.

Although no one was that surprised when Canterbury wasn't shortlisted for European Capital of Culture, local culture carried on regardless. When a new council crackdown on begging led to a legitimate busker (a double bass player) getting arrested under an obscure 19th century bylaw, a lively crowd of buskers assembled to protest outside the Police Station. He was eventually released and a discussion about possibly introducing licensing or some kind of charter for buskers began. The buskers weren't impressed – silent protests were staged in the Buttermarket. The Sidney Cooper Centre in St. Peter's Street, given to the city by the famous local artist to be an art academy, having fallen into decline and then come into the hands of Christchurch College, got a major revamp in 2004 and reopened as the Sidney Cooper Gallery. Over on Stour Street, the Canterbury Heritage Museum opened an adjacent Rupert Bear Museum in 2003, bankrolled by a National Lottery grant, to honour the creation of Canterbury native (and student of Cooper's academy) Mary Tourtel. Some months later, the Lord Mayor opened a Bagpuss mini-museum next to it, also featuring displays about Oliver Postgate and Peter Firmin's other beloved children's television creations. A twice-annual St. Dunstan's Street Festival took off in September 2003, becoming a great success, with bands playing in the street (Roper Road resident and key Canterbury scenester Richard Sinclair played at the second one, appropriately). As well as live music there were activities for kids, circus performers, street traders and face-painting. After three successful years, the Canterbury Fayre, a festival held at nearby Mount Ephraim Gardens, came to an end in 2004 when the company went bust. Leaning towards psychedelic and progressive rock, the Fayres had seen performances from Caravan, Gong, Hawkwind, King Crimson, Robert Plant, Love, Arthur Brown and more.

Some years earlier there'd been some talk of using it for a contemporary arts centre, then for a railway heritage museum, but in the end the old goods shed near Canterbury West station was turned into a farmers market, opening at the end of July 2002. The city finally got a mosque, too, after many years of applications being rejected. The Markaz, a house on Giles Lane directly across from the old Hopper family home Tanglewood, was to be converted into a space that could accommodate up to 140 worshippers. The city's Muslim community had acquired the property over twenty years earlier from UKC but attempts to secure permission for a mosque had been repeatedly blocked. By November 2004 building work had begun on the now cleared site of the old St. Mildred's Tannery.

11th December 2004 to 29th November 2006

The £140m Whitefriars shopping development officially opened in early September 2005 with a noisy fireworks display. Not everyone was impressed, the complete lack of greenery being a major complaint. Whitefriars manager Peter Scutt failed to win over the locals when he accused the nearby St. George's Street market of being "tatty", "cluttered" and "of appalling appearance". In a rather transparent attempt to blend in to the community, a "Whitefriars Street Fest" was staged during the August 2006 bank holiday weekend with street art, a costume pageant, live music and breakdancers. The £1000 prize for the busking competition was split between a pair of young singer-songwriters, UKC graduate Lucy Kitt and Liam Magill of the recently formed psychedelic-progressive rock band Syd Arthur (he, his brother Joel and a couple of their St. Edmund's school friends).

Over on the newly rebranded "King's Mile" the trial pedestrianisation of Palace Street was dropped in July 2005, the police having found it to be unenforceable. The newly formed King's Mile Traders Association, though, claimed that they had the support of three-quarters of traders for full pedestrianisation. A compromise was eventually reached with the pavements to be widened considerably, halving the width of the street and allowing for tree-lined pavements and the sought after "continental-style cafe culture". Pedestrianisation in the area around the Westgate was also being discussed. A trial six-month period of traffic-free Sundays began in August 2005. Claims were made that the 14th century towers were in danger of being destroyed by the relentless traffic. A petition with 1000 signatures was delivered to the council, urging them to reduce the 16,000 vehicles a day that were passing between or around the Towers.

On the night of 21st January 2005 another iconic feature of the city, the famous lime tree at the St. Lawrence cricket ground, was blown over in high winds. It had stood for 200 years with special rules introduced for balls hit into and over it, the only such anomaly in the cricket world. The tree had been known to be diseased, so a replacement had been planted nearby in 1999. The fallen tree was cut up and used to produce souvenirs.

Festival culture was on the rise. The twice-yearly St. Dunstan's Street Festival continued, with the May 2005 edition serving as the opening for the two-week Canterbury People's Festival. Local bands played to a happy crowd on a sunny day, alternating with comedians. The People's Festival continued at various locations around the city. The next Street Festival was a washout but the Falstaff Hotel opened its doors to storytellers and Orange Street Music Club put on live music all day. OSMC had recently taken over from Local Hero Records, a record shop and private member club/venue, having secured an extension of its limited music license. In 2006 it underwent a major refurbishment and reopened to the general public rather than just members. An idea which had begun at a 2003 St. Dunstan's Street Festival came to fruition in July 2006. Sean Baker of the local Lounge DJ crew was approached by farmer Tim Hume offering the use of his farm for "something a bit bigger". "Lounge on the Farm" was a two-day festival held at Merton Farm on the Nackington Road, featuring The Egg, Billy Childish, Nizlopi and local bands including Syd Arthur. It sold 2000 tickets, benefited from beautiful weather and made a lot of people very happy indeed. Canterbury also saw two gay pride parades in 2005 and 2006. Beginning at the Westgate Gardens, they passed through the city centre to the Dane John Gardens for a "Picnic in the Park". The second year featured a Wild West theme, enjoyed fine weather and even saw a couple of councillors join in. Despite being seen as a great success, with positive media coverage and the cooperation of the authorities, the event wasn't to happen again until 2016 due to a supposed "lack of support".

In April 2006 I visited Canterbury for the first time in a while, travelling from Exeter with my partner Vicky. As well as experiencing local singer-songwriter and raconteur Luke Smith's Thursday evening residency at OSMC, we visited Bigbury Hillfort and the Abbey ruins (my first time at either site), noting in particular the anomalous stones originally built into the foundations of St. Pancras Chapel and believed to have been recycled standing stones. On display in the Abbey visitors' centre were life-sized bronze statues of King Ethelbert and Queen Bertha, commissioned by sculptor Stephen Melton. A few weeks later, on St. Augustine's Day, Prince Michael of Kent unveiled the statue of Ethelbert on Lady Wootton's Green near the Abbey's Fyndon Gate, while the French Consul General unveiled Bertha. Buildings overlooking the Green now belonged to the ever-expanding Christchurch University – the once humble teacher training college founded in 1962 had been granted university status the previous autumn and rebranded itself as "CCCU". Members of its music department played medieval music for the unveiling ceremony, something of an anachronism as Bertha and Ethelbert had lived rather earlier, in the late 6th century.

There was talk of a license being granted for regular live music and dancing in the Dane John Gardens but local residents resisted this. Similarly, after an unnecessarily loud concert at St. Lawrence cricket grounds – Elton John playing to about 17,000 people – Kent Country Cricket Club sought a music and entertainment license but met similar resistance. That concert was the biggest one held in the city since The Clash rocked The Odeon in 1980 and many people enjoyed it from the comfort of their own gardens – but not everyone within a mile radius was entirely enthusiastic about this. Concerns were also being voiced about the Dane John Gardens, Westgate Gardens and St. Dunstan's churchyard becoming focuses of "anti-social behaviour", mostly involving alcohol and drugs. By late 2006 a significant increase in homelessness was becoming visible in the city. A group of men were reported to have been sleeping rough in one of the subways under St. George's Roundabout for weeks. Homeless organisations warned of a crisis and held an urgent meeting with the council and police. The council's response was to issue a blunt (and hollow) warning that street vagrants with no connections to the city would be bought train tickets and sent back to their home towns.

In March 2005 town planning consultants Yellow Book submitted a report to the council. *Canterbury: City of Imagination* attacked the quality of the shops, restaurants and hotels, as well as the lack of a "must-see area". Canterbury, it concluded, was not living up to its potential. Among its recommendations were investment in the Marlowe Theatre and Beaney Institute. By that autumn the council were seriously discussing a £24m rebuild of the Marlowe and the next spring began discussing major changes to the Beaney. They'd also announced that rubbish in black bags would no longer be collected, only in specially issued purple bags or wheelie-bins.

29th November 2006 to 22nd June 2008

At this time the city had three "Park & Ride" carparks: on the Sturry Road, the New Dover Road and in Wincheap. The possible siting of a fourth led to organised resistance from residents near two proposed locations: in Harbledown, on No Man's Orchard beneath Bigbury Hillfort; and in Thanington, on the recreation ground.

The rapid supplanting of postal by electronic communication led to the inevitable: after more than a century in business, the main Post Office on the High Street was closed and replaced by a much smaller Post Office in the upstairs of WH Smiths on St. George's Street. Many locals were troubled by this and shortly afterwards an anonymous street artist plastered stencils on the old building with an image of Oliver Twist asking "*PLEASE, SIR, CAN WE HAVE OUR POST OFFICE BACK?*" The small Post Offices at Oaten Hill, St. Dunstan's, Northgate and Wincheap also closed during this time.

Three years after Poland joined the EU there were now over a million Poles settled in the UK. Because of its convenient location, thousands were believed to be living in and around Canterbury. As indications of this, St. Thomas' Roman Catholic church began to hold a monthly mass in Polish and round the corner in Burgate Lane a shop called "Eastern European Food" was busy trading in sausages and Polish language newspapers. The Polish influx was described as a friendly one in the local press, which joked that Polish could almost be considered the city's second language.

A new community radio station went live in January 2007. CSR FM had received a five-year license and was aimed at 14-25 year olds. This subsumed two existing student radio stations, one of which, UKC Radio, used to host my late night eclectic programmes from '89-'94 (although I have no evidence that anyone ever listened to any of them). It was announced that 2007's summer carnival would be the last and would not feature the usual "Miss Canterbury" competition due to rapidly declining interest in the event. The carnival's time had passed and it was no longer relevant to young people – it couldn't even raise enough money to pay for its own insurance. But the sixth St. Dunstan's Street Festival went ahead early that summer. Despite the cold and heavy rain, the street was "a colourful sea of bouncing umbrellas". The second Lounge on the Farm festival at Merton Farm had much better luck with the weather, with 2000 festival-goers enjoying 160 acts (including Super Furry Animals, The Bees and Billy Childish) over a sunny weekend. I'd moved back to the Canterbury area that spring and was impressed by a new addition to the site: the Furthur tent, featuring a range of psychedelic and progressive bands. This was put together by half of the band Syd Arthur and some of their friends from the local outdoor free party scene, taking its (intentionally misspelled) name from the Merry Pranksters' legendary psychedelic bus made famous by Tom Wolfe's *The Electric Kool-Aid Acid Test* (1968).

Orange Street Music Club continued as a centre of local music culture but was fighting a battle with neighbours about licensing and soundproofing. By that autumn I found myself regularly spending evenings up at OSMC, sunk into one of their old sofas, listening to music and scribbling in a notebook, appreciating the relaxed, welcoming atmosphere. When the £600K makeover of the recently branded King's Mile was unveiled in October 2007 after months of roadworks, a street party was held with OSMC running the live music stage. On the more avant-garde front, the CCCU-sponsored Sounds New Festival took place in April 2008, one highlight being pioneering free-jazz saxophonist Evan Parker joining in with the dawn chorus in Blean Woods at 5 a.m., exploring the acoustics of the forest.

It cost the taxpayers £355,000 but was one of the biggest things to happen in the city for quite some time: in July 2007 an estimated 30,000 spectators turned up to watch the end of a stage of the Tour de France, which that year had an initial English leg. The race finished on the Rheims Way, appropriately (Rheims and Canterbury were twinned in 1962). Although no trouble was reported, the policing operation was the biggest thing to have been planned in Kent Police's 150-year history. The Australian cyclist Robbie McEwen, winner of the race's first stage, claimed it was the most people he'd ever seen at a cycle race. On some of the banners and flags that decorated the city for the occasion was the new slogan "Canterbury: Simply Inspirational". It's not clear who was responsible for this (I would guess the council paid far too much to a branding consultant) but it led to some derision. Professor Richard Scase took up the theme in the *Kentish Gazette*, attacking the city's uninspiring architecture. Less than a year earlier, property writer Max Davidson had described Whitefriars in the *Telegraph* as "*a text book example of 21st century Britain at its most mercenary...You don't need to be a Christian to be struck by the contrast between the nobility of the Cathedral and the tawdriness of the new shopping malls, which could be anywhere in England, they are so bland and charmless.*" Anthony Swaine, a respected local conservation architect still working at the age of 93, concurred: "*I couldn't have put the criticism better...an awful pastiche which, sadly, will not improve with time.*"

After repeated complaints by nearby residents about various types of antisocial behaviour in the Dane John and Westgate Gardens, a dedicated, uniformed park keeper was appointed to look after the two spaces. Police patrols were introduced in St. Gregory's churchyard after discarded needles were found there in the summer of 2007 and by that November special bins were placed in public toilets for the disposal of needles. The next spring Fenwick department store introduced a charge to use their toilets in an effort to deter heroin users.

After a spate of violent incidents, the Argyll and Sutherland Highlanders, based at Howe Barracks on Littlebourne Road, were banned from a number of local pubs and clubs, some displaying "no squaddies, no Scots" signs. Claiming unfair collective punishment and seeking to rebuild their public image, a summit was held between licensing officers, Army personnel and landlords, leading to a temporary "truce". In the autumn of 2007, the council closed the St. George's subway as its (rather typical) way of dealing with the fact that it was being used by homeless people as a place to sleep. This decision was based on claims that council cleaners were being attacked and abused. In January 2008 one of these cleaners reported that the subway was now just a "forgotten rubbish tip" and had been cleaner when the homeless people were sleeping there.

left: the new Sainsbury's supermarket in the Kingsmead area with its award-winning (but not universally loved) architecture; *right*: the 17th century ivory Virgin Mary statue stolen from the Cathedral Crypt on Epiphany 1981

left: the Odeon (formerly the Friars) cinema, also a music venue which hosted numerous punk and new wave bands in the late 70s and early 80s before being converted into the new Marlowe Theatre; *right*: the controversial (in its day) Marlowe memorial sculpture, having found a home outside the Marlowe Theatre after being moved from the Buttermarket to King Street to the Dane John Gardens

left: A sculpture by Mark Fuller, embodying the wind, one of three made from the root balls of large trees blown over beside The Causeway in the Great Storm of 1987; *right*: a new sculpture of Mary and Jesus created by the nun Mother Concordia (a.k.a. Carolyn Scott) and installed in the chapel of Our Lady Undercroft in 1982 to replace the stolen one

left: the awkwardly positioned tomb of Huguenot Cardinal Coligny near the former site of Becket's shrine, claimed by some to be the hiding place of Becket's bones; *right*: the limited edition teddy bear produced by a Christian website to celebrate the enthronement of Rowan Williams as the 104th Archbishop of Canterbury

left: the larger of the two mysterious standing stones in the ruins of St. Augustine's Abbey; *right*: one of a number of stenciled posters which were plastered on the city's main Post Office in 2008 when it was closed

left: Stephen Melton's sculptures of Queen Bertha and King Ethelbert, installed on Lady Wootton's Green (formerly "Mulberry Tree Green") on St. Augustine's Day 2006; *right*: the Bertha sculpture, with the Abbey's Fyndon Gate in the background (at this time an entrance to King's School premises)

22nd June 2008 to 22nd September 2009

In August 2008 I set foot in the second Marlowe Theatre for the last time (the first time had been to see Fairport Convention as a student in late '88). The Senegalese band Orchestra Baobab were playing as part of that year's Canterbury Festival. Quite a few of us in the audience ended up dancing in the aisles to one of the most joyful and beautiful outpourings of music I've ever witnessed. By the following June, demolition work was underway, the seventy-year-old building to be replaced by a modern £25m edifice (with local celebrity Orlando Bloom being made its patron). Like the £11m the council approved for an extension to the Beaney Institute that summer, the new Marlowe was an idea which had been conceived some time before the financial crash of 2008, otherwise it almost certainly wouldn't have happened. The "credit crunch" hit Canterbury, manifesting in a surge in homelessness, council budget cuts, increased Council Tax, numerous businesses closing and a drop in traffic on the roads (accompanied by record sales being reported by local bike shops). The council had £6m tied up in crisis-hit Icelandic banks which meant that plans to transform Wincheap and improve the city's traffic situation were put on indefinite hold.

Despite the economic climate, culture in Canterbury continued its pleasing evolution. Lounge On The Farm 2008 saw 5000 people enjoying 160 acts (including the New York Dolls and The Coral) during another hot sunny weekend and a few weeks later the organisers opened their own permanent venue, The Farmhouse, in Dover Street. I made it along to the 2009 LOTF to see a version of the classic Gong lineup playing on the now expanded Furthur stage. In a BBC Kent interview, guitarist Steve Hillage mentioned having gone to have a look at his old student lodging in St. Radigund's Street. Soft Machine bassist Hugh Hopper had died earlier that summer and Daevid Allen began Gong's magnificent set by dedicating it to him, a dear friend and musician he'd first played with in 1963. Emerging local psychedelic rock band Syd Arthur also paid tribute with a live version of Hugh's *magnum opus* "Facelift". A group of teenage Canterbury Scene enthusiasts, The Boot Lagoon, asked Daevid to come and watch their set earlier in the day and he later spoke enthusiastically to the BBC about "the Canterbury Muse" singing once more.

The "Canterbury Muse" was further in evidence with the growth of a vibrant local live music scene. Along with The Farmhouse, Orange Street Music Club continued to promote this, as well as bringing in a succession of acts from elsewhere. By this time I was frequently attending gigs, regularly seeing local acts such as Syd Arthur, Zoo For You, Sávlön and the Jimmy Jones Band. Of particular interest to me was a monthly acoustic "Moonlit Fingertips" night at OSMC, hosted by Liam Magill and Raven Bush from Syd Arthur whose playing (both acoustic and electric) was rapidly maturing into something of profound beauty.

August 2008 saw the first instance of an annual "Goddess Parade" down the high street, linked to a "mystic fair" being held in the Westgate Hall: a curious melange of pagan women in Goddess and witch outfits, Morris dancers, neo-druids, belly dancers and little girls dressed as fairies. That Easter also saw the first "Magic Faery Festival" parade. Behaviours and beliefs that in earlier centuries would have led to persecution and possibly death were now being cheerfully celebrated in the local paper. The *Gazette* also came out against the council's attempt to tightly regulate busking in the city centre, a hare-brained scheme involving permits, one-hour busking limits, only six approved spots and fines of up to £20,000!

On the evening of 31st August 2008 dozens of people reported seeing a UFO near St. Stephen's Hill, the display involving pulsating multicoloured lights and sudden, rapid, acrobatic motions. Documents released in 2013 confirmed that governmental authorities were aware of this but claimed to have no idea what it was.

The demographic profile of the city was changing rapidly. A survey of foreign children in district schools revealed that at least seventy-four languages were being spoken. St. Stephen's Juniors had the highest number of kids whose first language wasn't English, including Afrikaans, Akan, Albanian, Amharic, Arabic, Armenian, Bengali, Cantonese, Hakka, French, German, Greek, Gujarati, Hebrew, Hungarian, Italian, Japanese, Malayalam, Malay, Ndebele, Polish, Persian, Romanian, Shona, Spanish, Tagalog, Urdu and Yoruba.

Changing student attitudes were also becoming evident. Rather than the occupations and protests of the 60s, 70s and 80s, Kent University spawned a Facebook group in early 2009 campaigning for a Primark outlet to *open* in Canterbury (elsewhere in the country, some student groups were picketing the chain for its links to sweatshops and child labour practices). More encouragingly, students from UKC's new Conservation Society were busy planting a wildflower meadow and native species of trees, maintaining ponds and a nature trail, and coppicing glades in an area of woodland on the campus.

A curious new addition to the UKC campus was a traditional pre-Christian labyrinth made of Portland stone and set into the hillside below Eliot College, aligned with the Cathedral. Its seven-circuit design displaying the medieval "four fold" style, this was opened in October 2008 but a sneak preview had been given earlier in the year during the 14th Lambeth Conference. That Conference was presided over by Archbishop Rowan Williams who was doing his best to prevent the proceedings from being dominated by the issue of gay bishops as they had been ten years earlier. In September 2009 CCCU's new £35m library complex, Augustine House, opened on the Rheims Way. Vice Chancellor Professor Michael Wright presided over the ceremony, declaring that Christchurch was now on an equal footing with its rival up on the hill, "no longer the little brother to Kent [University]".

Also in September 2009 the first high-speed train took commuters from Canterbury West to London St. Pancras in just over an hour with the Lord Mayor and MP Julian Brazier on board. Tickets were quite a bit more expensive but the shorter journey time instantly made Canterbury more attractive to commuters working in London, shifting the economy and demographics of the city in that direction. Meanwhile, a stone's throw from Canterbury East station, the Canterbury Open Centre homeless shelter had to close its doors to new arrivals due to a funding crisis in the associated Scrine Foundation.

22nd September 2009 to 19th September 2010

In early October 2009 Cathedral stonemasons held a "topping out" ceremony to mark three years of work finished on the Corona Tower (this had been built in late 12th century to house the severed crown of Thomas Becket's skull). The next summer, another topping out ceremony was held at the new Marlowe Theatre as its pinnacle was lowered into place. Council leader John Gilbey addressed a crowd from the top of the theatre, declaring this to be a major event in the history of Canterbury. Prior to construction work, archaeologists working on the site of the old Marlowe had discovered the remains of a Roman townhouse. An archaeological dig was also going on at the back of the Beaney Institute prior to its extension being built, discoveries including an extensive network of small shops and lanes from Roman Canterbury.

Graffiti again featured prominently in the local press when politics, philosophy and economics student Kitt Klarrenberg had his Spring Lane flat raided and was linked to the prolific "OREO" tag which had been appearing around the city. He was fined £200, causing some consternation, as the council claimed to have spent £66,000 cleaning graffiti in 2009. Meanwhile, the local graffiti art community expressed outrage when award-winning artist Tom Stanley was given an eighteen-month sentence for graffiti damage caused in 2005-2008 which the council claimed amounted to £52,000 (that would have been about £4000 per piece). Pointing out the disproportionate severity of his sentence compared to more serious criminals, a Facebook group was created to campaign for his release.

Council spending cuts became the target of protests with a group of about a dozen protestors disrupting a council meeting in early October 2009. Irresponsible, risky investments in Icelandic banks were decried: "now ordinary workers and local taxpayers are going to suffer". To save £3.5m the council had decided to close the Roman Museum and Westgate Towers Museum for part of each year, discontinue the traditional Christmas lights and (once again) threaten to demolish Westgate Hall. By the next February a thousand residents had joined the Facebook group "Save Canterbury's Museums". An angry protest was held in Butchery Lane, attracting hundreds despite the snow, many of those involved heading across town afterwards to join a protest against the closure of Westgate Hall. That summer, the Westgate Towers Museum was saved by entrepreneur Charles Lambie who announced his intention to invest £250,000 in it. Around the same time, the newly formed Westgate Community Trust put together a business plan to save the Hall, holding a street party nearby with live music, dancing, comedy, food and children's entertainment.

Large numbers of students moving through the city attracted news attention on two occasions for very different reasons. October 2009 saw 600 students on a "Carnage" pub crawl whose screaming and swearing caused some alarm among the cultured citizens emerging from a St. Petersburg Radio Orchestra concert in the Cathedral. The event organisers "Varsity Leisure", it transpired, had gone as far as bussing in students from Kent University's Medway campus. In March 2010, united students from UKC, CCCU, Canterbury College and UCA (formerly KIAD, the art college) marched against increased tuition fees, chanting "love Canterbury, hate debt".

The "Canterbury City Council Bill" was passed in the House of Commons in July 2010, finally allowing the local council to bypass national street trading laws and restrict the number of barrow traders operating in the high street. MP Julian Brazier had argued that Canterbury was now the third or fourth most visited city in the UK but had medieval-width streets in places. Visitor numbers and traffic remained sources of local concern. The new St. John's Coach Park in Northgate was opened with a ribbon-cutting ceremony by the Lord Mayor in February 2010 but it was far from a cause for local celebration. With places for forty-five tourist coaches it had required the felling of dozens of beautiful, mature plane trees. Protests continued against No Man's Orchard near Harbledown becoming the next Park & Ride site, although everyone agreed that *something* needed to be done about traffic in the city. Articulated lorries trying to avoid congestion were getting stuck in the back streets of Wincheap. A £175,000 cycle path was built along the Stour between the Westgate Gardens and Chartham via Hambrook Marshes, a move that was largely praised, although (typically) some decried it as a waste of money. A footbridge and lift was finally installed at Canterbury West station in the autumn on 2010.

Having been back in the area for three years, I found myself getting increasingly immersed in the burgeoning local music scene, a profusion of events now happening at Orange Street Music Club, the Farmhouse on Dover Street and Caseys on Butchery Lane. Syd Arthur were becoming particularly impressive. As well as numerous full band performances, I'd caught a number of acoustic duo sets from guitarist Liam and mandolin/violinist Raven as part of their "Moonlit Fingertips" nights at OSMC and a wondrous freeform jam on the evening of the 2009 winter solstice at Caseys. I was very happy, then, when they agreed to play for my 40th birthday party which I hosted at Littlehall Pinetum, along with Cocos Lovers, a more folkie/rootsy crew from Deal who were by now regularly gigging in the area. This cemented an emerging friendship with both bands and later that summer I found myself helping out members of Syd Arthur and the associated Furthur collective in setting up their zone at Lounge On The Farm. This was to be the last year Furthur hosted their stage at LOTF but it was a great success, with acts including Led Bib, Polar Bear, The Moulettes, Quantic, Circulus, Wolf People, School of Imagination, The Boot Lagoon, Zoo For You, Happy Accidents and Jah Wobble.

Various social connections were emerging through the music. When "punk jazz" quartet Led Bib played OSMC in January 2010 I met several uncharacteristically interesting UKC students, all members of the newly formed UKC Psychedelics Society. I'd spotted a (thoughtful and sober) poster for this in the campus library earlier that day and reflected on how such a thing would have been unimaginable back when I'd been a student there. That August at a farmhouse party near Boughton, as well as catching an impressive set from prog-rockers The Boot Lagoon and a jam involving them with members of Zoo For You and Syd Arthur, outside by the fire I met Rob Gambell, a local singer-songwriter (with a punk background) and aspiring ecclesiastical stonemason with a lot to say about the history of Canterbury and its religious establishment. That evening I also got to know Ed S, one third of "Ed, Will and Ginger", three young nomadic folksingers from the area who'd been walking the country lanes, public footpaths and ancient trackways of Britain for some time now, learning songs along the way and literally singing for their supper.

We're now working with less than a year per episode – at this point we'll take them down to single pages.

19th September 2010 to 6th July 2011

In December 2010 the University of Kent announced plans to build new accommodation units for 800 students plus a conference centre on the green slopes between the campus and town (an area which had suddenly, mysteriously acquired the name "Chaucer Fields"). This led to immediate organised resistance from residents' groups, with some support from ecologically-minded students. The protest attempted to secure "Village Green" status for the site as well as focussing on the protection of its ancient hedgerows.

Students were being hit with massively increased tuition fees so some student activists occupied the UKC Senate building over Christmas and New Year. Similar occupations occurred around the UK at this time but this one lasted the longest, with a hard core of half a dozen occupiers braving freezing temperatures, leaving voluntarily to the applause of supporters on January 5th. Spokesperson Alan Bolwell, incidentally, was once a housemate of Kitt "Oreo" Klarrenberg, the prolific (and convicted) graffiti writer.

The local music scene suffered a loss when Orange Street Music Club closed in October 2010. I helped the Furthur collective redecorate the space for one last blow-out featuring Parisian Afrobeat band Cafe Creme and local "Afro-Kentish" crew Mr. Lovebucket. A few months later, the space had been reopened as The Ballroom, with much more emphasis on alcohol sales and less on music. Members of Syd Arthur set up Dawn Chorus, an independent record label for the local scene, and during this time I ended up at several jam parties at a bungalow on the New Dover Road, from where a new maximalist prog band called Lapis Lazuli were starting to operate.

In April 2011 an interdisciplinary conference on psychedelic consciousness took place at UKC's recently opened Woolf College over three days. The brainchild of several members of the UKC Psychedelics Society, "Breaking Convention" was an undoubted success, featuring eighty speakers and 600 attendees from thirty countries with discussions touching on archaeology, history, literature, philosophy, medicine, anthropology, law and politics. For financial reasons (rather than controversy) the next three biannual Breaking Conventions were held at Greenwich University, arguably lacking the groundbreaking spirit of the first one.

Another exciting splash in the local cultural pond occurred a few weeks later with the Sondryfolk collective's "A Canterbury Trail", an event which involved performances and art installations around the city centre one Saturday in late May. I'd met Elise and Laurie, two thirds of Sondryfolk, via the Furthur crew at LOTF the previous summer and enthusiastically volunteered to help out with what turned out to be a thoroughly inspiring occasion. Among the locations used were the undercroft of the Eastbridge Hospital of St. Thomas, the Westgate Gardens (with an aerialist performer in the huge old plane tree) and the Greyfriars Garden tucked away off Stour Street, the latter which I'd failed to notice in all my years around Canterbury. The event concluded with feasting and live music in the tiny Best Lane Garden, somewhere I'd often passed but similarly failed to really notice. My overriding memory is of a blackbird singing while Liam and Raven from Syd Arthur played beautiful acoustic duets in the evening sun.

6th July 2011 to 21st February 2012

Now in its sixth year, Lounge on the Farm took place 8th-10th July, with a much expanded capacity of 10,000. There was no longer a Furthur Field, on-site crime was up 50% and there was a general consensus that the event was losing touch with its roots. I certainlly felt that, but helped out with the new Meadows Field and enjoyed seeing local post-rock band Delta Sleep (formerly Sávlön) open its stage on the Friday afternoon. I saw them again at a party in August, one of several bands playing in the garden of Lapis Lazuli's shared house at the confluence of the Old and New Dover Roads. That October I was again involved with the Sondryfolk collective, this time helping to organise an "autumnal jamboree" – an ambitious free outdoor event involving art, music, performance, feasting and mini-lectures.

January 2012 saw the launch of Free Range, a weekly series of free avant-garde music, poetry and film nights at the Veg Box Cafe above Canterbury Wholefoods on Jewry Lane. If they, and the Sondryfolk event, represented the artistic underground of the city, the official face of culture was the new £25.6m Marlowe Theatre which had opened with much fanfare in early October 2011. The first major event to take place was the annual pantomime, this time an adaptation of *Cinderella*, featuring the usual cast of minor TV celebrities, but for the first time in many years *not* featuring Dave Lee, the well-loved local comic, pantomime "dame" and charity fundraiser, who'd pulled out due to illness. He died of pancreatic cancer in January, aged 64, his local popularity being reflected by the memorial service held for him in the Cathedral.

Another high-profile death occurred that January. Charles Lambie, the entrepreneur who'd saved the Westgate Towers museum by investing heavily in the property was found dead from shotgun wounds at his home in London, a suicide brought on by depression precipitated by financial problems. A month later, a double tragedy hit the city when the ubiquitous busker Daniel "Taihg" Lloyd and his young beatboxer friend Hugo Wenn were found drowned in Reed Pond near the Howe Barracks. An outpouring of grief and love was evident in the streets of the city centre when Taihg's favoured busking spots were inundated with flowers, photos and chalked tributes which accumulated over several weeks, after the city's busking community gathered in the high street for a wake/jam in his honour.

The campaign to save Chaucer Fields from an expanding UKC campus continued with a formal notice of application for Village Green status being made to the University including 360 witness statements about recreational use of the space going back decades. Meanwhile, another green space had come under threat, the council now considering the sale of Kingsmead Field to property developers. Yet another campaign which began at this time was to save the Kent and Canterbury Hospital's maternity unit. The authorities were threatening to close this, meaning that women from Canterbury would have to travel to Ashford to give birth. A campaigning success came in August 2011 when the Westgate Hall Trust entered into an arrangement with Curzon Cinemas to turn part of the property into an arthouse cinema, thereby saving the community hall from the council's (characteristically unimaginative) plans for a car park.

21st February 2012 to 23rd August 2012

After several weeks, Daniel "Taihg" Lloyd's family asked for floral tributes and pictures to be removed from the high street corner where he used to busk. A colourful memorial service was held in late February at St. Peter's Church with 200 mourners celebrating his life and music. A few weeks later a toxicology report revealed that he and his friend Hugo had been under the influence of the "legal high" MXE when they drowned. This sparked a campaign (involving Hugo's parents, MP Julian Brazier and former archbishop Lord Carey) to close the Skunkworks headshop in Northgate which was selling MXE and other "research chemicals".

In April a campaign to save Kingsmead Field was launched. The council had claimed that it had no choice but to build on greenfield land around the city but a petition of 1300 names forced a debate on the matter. Meanwhile, UKC had announced that its plans to build on Chaucer Fields were on hold due to the application which had been filed for Village Green status. The campaign to save the K&C's birthing unit soon conceded defeat, though, with Kent County Council rubberstamping the decision to close it in early June.

The 1011 Viking siege and 1012 martyrdom of Archbishop Alphege were commemorated by the planting of a Judas tree beside St. Radigund's Bridge on April 21st. Divergent attitudes to the future appearance of the city were evident in the rejection of the Abbot's Mill Project's plan to install a bank of solar panels at a location nearby. Led by a couple from Boughton, the Project had been pushing to develop the former mill site with a "sustainable" community centre and hydroelectric waterwheel, but various residents' associations argued that the panels would simply be an eyesore. Likewise, the revamped St. Lawrence cricket ground attracted much disapproval, particularly the inclusion of a small Sainsbury's supermarket.

The biggest source of local grievance in many years took root in late March when a long-discussed traffic scheme was finally imposed by the council on the Westgate area. Supposedly intended to safeguard the historic towers and reduce air pollution in the notoriously congested St. Dunstan's Street, the result was a marked worsening in traffic conditions leading to widescale local outrage and a vigorous campaign to reverse the scheme.

In June, a beautifully composed and executed life-sized stencil of local music legend Robert Wyatt (in his wheelchair, pensively smoking a cigarette) appeared overnight on a wall on Dover Street next to the site of the old Beehive club where his first band, The Wilde Flowers, had regularly played in the mid-60s. Rather than vandalism, the *Gazette* described the piece (signed "Stewy") as a "mystery drawing of [a] music legend", reacting with curiosity rather than condemnation. The stencil was just across the road from The Farmhouse and I can vividly recall the aftermath of a Syd Arthur gig there that summer when people went over to have a closer look and have their photo taken with "Robert". The local music scene was as busy as ever. I was often going out to see bands like Syd, The Boot Lagoon, Lapis Lazuli and Arlet, the latter a magnificent new "chamber folk" quintet who'd played their first gig in the undercroft of Eastbridge Hospital as part of the Sondryfolk arts trail the previous spring.

23rd August 2012 to 17th January 2013

After a three-year, £14m refurbishment, the city's Victorian library and museum reopened in September as the new, improved "Beaney House of Art and Knowledge". The dusty old display cabinets had been replaced with a much more modern form of curation, there was now a cafe and the whole place had a much brighter, more modern and inviting feel. As the new Beaney opened, The Farmhouse closed, resulting in the loss of twenty jobs and leaving Canterbury short of a dedicated music venue. But the local arts and music scene continued to thrive when the "Free Range" series of weekly avant-garde events entered its second season. As part of the season I was invited to give a "free-jazz maths lecture", backed up by three improvising musicians as I instructed an amused and bewildered audience in the workings of Euler's Formula $e^{i\pi} = -1$. That autumn also saw the culmination of a monthly podcast I'd been putting together: *Canterbury Soundwaves* had run for two-and-a-half years, exploring and celebrating the "Canterbury Scene" music of the 1960s and 70s. For the final episode I recorded an extended interview with Soft Machine founder Daevid Allen with whom I spent a memorable day wandering the city, chatting about fifty remarkable years of his creative life.

Controversy over the Westgate traffic scheme intensified during this time with the "Get Canterbury Moving" campaign holding public meetings in which hundreds of residents expressed their outrage. By November it had become abundantly clear that the scheme wasn't working, polls suggesting 80% local opposition, but the council vowed that it was to become permanent. In November, though, Kent County Council overrode Canterbury City Council and decided to pull the plug in March 2013 when the one-year mark was reached, returning the road system to how it had been.

The Save Chaucer Fields campaign seemed to be having some success as UKC changed its proposed construction plans, offering a "compromise" plan. This was swiftly rejected by protestors who presented their Village Green application to KCC officials at the newly saved Westgate Community Hall. Things didn't look so good for the campaign to save Kingsmead Field, though, as the council voted to seize control of it despite powerful local opposition. This sparked claims of undemocratic governance in the district. Growing levels of poverty in the district led to the setting up of a Canterbury Food Bank, part of a growing phenomenon around the UK since the 2008 financial crash. It operated from a double garage on Wincheap retail park and was linked to the evangelical Canterbury Vineyard Church.

A new Archbishop of Canterbury was selected during this time: Justin Welby, a former oil executive and (briefly) Bishop of Durham. An ancient ceremony was held to formally elect him in the Cathedral Chapterhouse in January involving the "College of Canons of Canterbury", some members flying in from Rome and Zürich for the occasion. National Census results published during this time showed that 60% of local residents considered themselves Christians, down from 73% in the previous census and the lowest in any district in Kent. The population was also revealed to contain 724 adherents of the Jedi faith, thirteen Scientologists and one Satanist.

17th January 2013 to 14th May 2013

The disastrous Westgate traffic scheme ended in April despite the council's desperate attempts to keep it alive. In the meantime a variety of plans to tackle the problems caused by 21st century traffic flow in a small medieval city had been proposed: trams, tunnels, a bypass, car-sharing schemes, more cycle lanes, more busses... More opposition to the council's plans occurred when the Save Kingsmead Field Campaign mounted legal action to block them from selling the site off to developers.

New Archbishop Justin Welby was enthroned at the Cathedral on March 21st. Visiting Prime Minister David Cameron became the subject of a "Carnival Against the [public sector] Cuts" organised by local activists. 47-year-old Jonathon Elliott from Whitstable hurled himself at a Bentley carrying Prince Charles and his wife Camilla down Northgate, quickly being wrested away from the car by police. A couple of days later I was cycling back from a "Fruits of Spring" festival at Woolton Farm near Bekesbourne, following the old pilgrims' route back into the city. This took me down High Street St. Gregory's (a narrow lane formerly on the grounds of the medieval Priory of St. Gregory) along which I caught a whiff of pungent cannabis smoke. Looking up I saw a huge cloud of the stuff floating out of an open window in which was placed a recently appropriated yellow "No Parking"-style sign – not an uncommon student acquisition, except that this one said "*No Stopping or Loading: Enthronement of Archbishop*"!

Two sensational local crimes hit the national press: in early February, members of a multi-million pound cocaine-dealing gang were given a total of twenty-eight years; in early March, a 57-year-old nurse was caught up in transatlantic cannibalism allegations. The latter had been active on cannibalism web-forums, recently bought an axe and had expressed an unsavoury interest in a female Chinese student working at "Chop Chop", his local takeaway in St. Dunstan's.

The last petrol station in the city centre, a Total garage on the ring road, closed in early 2013. By April, its covered forecourt had become an ad hoc homeless shelter. As a result of some (allegedly Polish) residents lighting fires, though, the premises were soon enclosed by security fencing, forcing the homeless to look elsewhere for shelter. The city's prison, opened in 1806, closed at the end of February, its 314 inmates (all foreign) being moved elsewhere and most of its 183 workers being moved to other jobs in the prison sector.

The brilliant and eccentric Kevin Ayers (Wilde Flowers, Soft Machine) died in his sleep at his home in southern France in February. I got the news while on my way back from Bristol where I'd been to see Syd Arthur, their spring tour sadly not including any local gigs due to a lack of suitable venues. They did play in town in May as part of the "City Sound Project", a city-wide festival aimed at students and centred on alcohol consumption. Their gig was in the garden of the Jolly Sailor so I decided to sit at the end of St. Radigund's Street and listen from there, comfortable with a cushion, blanket and flask of wine. The blanket led quite a few passers-by to assume I was homeless, provoking a range of responses from derision to compassion – a very interesting evening.

14th May 2013 to 15th August 2013

In May the council caused much local disgruntlement by rolling out their new £1m "six bin" rubbish and recycling scheme. The new containers were seen as "confusing, unhygienic and inadequate". A few weeks later, binmen from Serco (who'd recently won the waste collection contract for the district) were caught on camera mixing recycling and rubbish before putting it in their lorry. Further claims of incompetence followed: food waste being left strewn on pavements or left uncollected for weeks. Further discontent was caused by the council's decisions to close its civic museums for several months each year and not to install Christmas lights in the city centre due to budget cuts. They came in for further criticism when a wooden bench placed on Kingsmead Field by campaigners was removed by Serco as a "health risk" and a beautiful 100-foot plane tree outside the Victoria Hotel on London Road was scheduled for felling because it was supposedly blocking a pavement. Due to the volume of angry responses the bench was replaced and the tree allowed to live. Also stirring up discontent was the council's new "Local Plan", a blueprint for Canterbury's future which involved plans for up to 4000 new homes to be built on farmland south of the city.

As Lounge on the Farm approached (again with a capacity of 10,000), fears were expressed by some locals of having a mini-Glastonbury on their hands. The 2013 lineup didn't impress a lot of people, though, with Soul II Soul, Aswad and Seasick Steve headlining. At the more experimental end of the artistic spectrum, I found myself helping to organise "Piano In The Woods", a cycle of thirteen monthly performances on a decaying outdoor piano in Littlehall Pinetum by Free Range events organiser Sam Bailey, collaborating with a diverse array of poets, musicians, singers and dancers. I also helped to facilitate an artists' residency on the same site for chamber folk ensemble Arlet who spent a few days camping and working on new music, culminating in an unforgettable fireside performance alongside former members of Penguin Cafe Orchestra.

Having been evicted from the former Total petrol station, several homeless Eastern Europeans set up camp in St. Mildred's churchyard. This caused some concern with neighbourhood residents but there was also some local alarm at a series of random attacks on homeless rough sleepers, three reporting having their sleeping bags set on fire. A charity that had been giving out free food and clothing in North Lane Car Park every Sunday for nine months was ordered to stop after complaints about "antisocial behaviour". The organiser, a former soldier, claimed that the council tolerated his work over winter but wanted to sweep the homeless problem under the carpet during the tourist season.

In late June, the 5th Battalion the Royal Regiment of Scotland (or "Argyll and Sutherland Highlanders") ended their ten-year stay at Howe Barracks which the MoD had decided to close. The Queen, their "Colonel-in-Chief", visited to witness their final parade and leaving ceremony. Howe Barracks was to continue to serve as the base for a battalion of reservists until its closure but a final parade of the "5 Scots" (a battalion of 500) through the Westgate and Cathedral Precincts with their Shetland pony mascot, pipers and drummers effectively marked the end of a military history in the city going back to the first barracks being built in 1793.

15th August 2013 to 28th October 2013

The traffic controversy raged on as council chiefs held secret meetings on new plans for the Westgate road system. Claims were made that bus company Stagecoach had pressured them into the disastrous scheme because a new fleet of buses they were planning to introduce would be too wide to fit through the Westgate but they wanted to continue operating their lucrative routes through St. Dunstan's. After the council refused to hold a public meeting on the matter, the *Gazette* organised their own at which the vast majority voted for the "do nothing" option for the traffic arrangements in the Westgate area. As expected, the council made Kingsmead Field available to developers on the supposed basis that it was hardly ever used. One councillor claimed he'd visited and only seen a crow and a seagull. Campaigners reacted with a "Crows v. Seagulls" cricket match on the field in early September, 160 people turning out to watch.

A couple of weeks later, Cathedral staff were mourning the death of a long-term resident, a much loved and affectionate cat, named "Laptop" by the choirboys when he'd mysteriously shown up there. He was laid to rest with floral tributes in the Water Tower Garden, his favourite spot for lounging in the sunshine. For years he'd mewed each morning at the Martyrdom door just before Matins so that he could attend the service, even being given his own chair.

In late October, someone in a Guy Fawkes "horror mask" was terrifying women late at night in the city. At least one was pounced on but fortunately escaped after punching him in the groin and running away. Police attempts to track him down were complicated by the approach of Hallowe'en, a similarly masked person accosted by police in Station Road West turning out to be a woman who had nothing to do with the incidents. Fear for women's safety had been cited as a cause for objection in the previous few months after the council had controversially agreed to grant a license to "The Bing", the city's only strip club, which opened in the old Farmhouse premises in early September.

The third Free Range season began on October 3rd with a set from the legendary musical experimentalists AMM (who had at one time included former Cathedral chorister Cornelius Cardew). Towards its end, dramatic fork lightning became visible in the distance through the Veg Box Cafe windows. By the time I was cycling home, a massive thunderstorm was passing overhead. The air felt completely electric and I was soon soaked to the skin despite wearing full waterproofs. Passing the Millers Arms, I spotted Juliet (someone I knew vaguely via the UKC Psychedelics Society) and her friend Indyah, equally soaked and in a state of what could only be described as meteorologically induced ecstasy. I stopped to babble excitedly with them about the documentary *Act of God* I'd watched the day before (about being struck by lightning), particularly postmodern novelist Paul Auster's account of seeing a friend struck as a teenager – he turned out to be one of Indyah's favourite authors. I pedalled off up St. Stephen's Hill, feeling the possibility of being struck to be quite real, then got completely lost and disoriented in a familiar grove of trees, deafening explosions of thunder overhead and the rain almost *solid*. Meanwhile, down by the Friends' Meeting House, Juliet and Indyah had decided that they couldn't get any wetter and so got into the Stour and conducted a freestyle wedding ceremony.

28th October 2013 to 26th December 2013

Local frustration with the council reached a peak when a petition of 3000 names was presented to it calling for the entire Tory executive to be removed, mainly for their handling of the Westgate traffic trial and their moves to sell off of Kingsmead Field. The fact that there were to be no Christmas lights again this year only added salt to the wound. A "Bah Humbug" protest march was staged in the high street in early December, a large banner reading "*Please Sir, can we have some more...Xmas lights?*". That they were planning to spend £800K revamping the Lord Mayor's parlour (Tower House in Westgate Gardens) and £17K on refurbishing his robes didn't go down well with the city's Independent Traders' Alliance or the Tax Payers' Alliance.

In November, several members of the council on their way to a meeting in the Guildhall walked past a woman slumped on a bench in the Westgate Gardens. She was suffering severe hypothermia and was saved from certain death when more conscientious public servants called paramedics. Councillor Ida Linfield lashed out at her fellow councillors for walking away: "*It's all very well holding sponsored sleep-outs for the homeless in Canterbury Cathedral, but why walk away when confronted with something like this?*" (A charity sleep-out, involving the Lord Mayor, Bishop of Dover and other local luminaries had been held in the Precincts in September.)

A spate of violent incidents occurred in the city during this time, including three stabbings in two weeks (on the Spring Lane Estate, London Road Estate and close to Canterbury East station). Late one night, two young Poles were caught on CCTV carrying out unprovoked pepper spray attacks on strangers in the high street. CCTV also captured some extreme police brutality: a young shoplifter being repeatedly punched in the face while pinned to a wall by his throat in Debenhams. *He* was initially charged with assault, the charges quickly being dropped when the footage emerged. Dozens of police officers and dogs swooped on the NatWest bank one afternoon in November when an Ashford man brandishing a knife and fake gun attempted to rob it, the street outside being sealed off and nearby shops evacuated.

Tributes poured in from Canterbury Scene enthusiasts worldwide when Richard Coughlan, Caravan's drummer since 1968, passed away. In more recent years he and his wife had been running pubs in the area, including The Cricketers in St. Peter's Street. The avant-garde spirit of the earliest Canterbury Scene creativity was living on with the Free Range events and the "Piano In The Woods" performance series. I took part in a Hallowe'en Free Range, again lecturing on mathematics as my eccentric alter-ego Professor Appleblossom with improvised musical accompaniment (this time my topic was the "Fischer-Griess Monster sporadic simple group"). The seventh PITW featured contributions from CCCU's improvising Scratch Orchestra and Choir, the piano by now in a *very* sorry state.

To the delight of local cyclists and eco-activists, the recently constructed Great Stour Way cycle path, having suddenly come under threat, was saved by an anonymous supporter: she offered to buy the fifty-acre Hambrook Marshes site from Kent Enterprise Trust who'd put it up for sale, potentially compromising the future of both the marshes and the path.

early 21st century

26th December 2013 to 11th February 2014

On January 25th, a thousand years of Cathedral tradition was broken when a girls' choir sang at Evensong. Despite having only had three rehearsals, they found themselves the subject of an international news story and received a standing ovation. A few days later, just back from a trip to New Zealand, I got to see one of the first live sets from Koloto at The Lady Luck. "Koloto" was the artist name adopted by local electronic producer and bass player (formerly of Sávlön/Delta Sleep) Maria Sullivan whose "organic glitch" style of electronica was attracting an extensive underground audience worldwide. Despite being jetlagged and ill, I was happy to be there, back among friends and experiencing some deeply creative, uplifting music.

Save Kingsmead Field Campaign members were cautiously jubilant when the council announced that they were going to spare the field from developers, although they weren't prepared to go as far as to grant it Village Green status for its long-term protection. Meanwhile, work was about to begin on converting the recently saved Westgate Hall into a community centre and cinema. The "Get Canterbury Moving" campaign group was disbanded, having achieved its aims and amidst fears that the group was becoming too political. The Canterbury Food Bank was seeing increasing activity with a reported 4000 children in the Canterbury district now living in poverty. The charities MyStreets and Porchlight launched a joint initiative involving homeless people giving tours on Saturday mornings, showing both the city's obvious landmarks and quirky details, but also giving some insight into what it's like to be homeless.

Having recently received a 3300-name petition calling for the mass resignation of its executive, Canterbury City Council allowed just fifteen minutes for debate on the issue. After voting 31-11 against the motion, angry protests erupted, leading the Lord Mayor's chauffeur to call the police. The campaign against undemocratic governance had been organised by businesswoman Debbie Barwick and during the council meeting the locks on the door of her vintage clothing shop Revivals in St. Peter's Street were superglued. Weeks later one of her shop windows was smashed, but as nearby businesses Pure Magik (crystals, incense, etc.) and Doo-Das (a hairdressers) were also vandalised on the same night, Ms. Barwick didn't believe the attacks to be related.

February saw major flooding, with a month's rain falling in just a few days. Nearby villages on the Little Stour were the worst hit but Canterbury also suffered, with flooded areas including St. Radigund's Car Park, Westgate Gardens, Sainsbury's car park and the Rheims Way underpass.

A student-oriented website was slammed for enthusiastically describing the mass consumption of "Jägerbombs" (a mixture of the spirit Jägermeister and the energy drink Red Bull) at the Old Brewery Tavern in Stour Street as "heroic" and "record-breaking". Nearly 5000 of the £1 drinks were consumed during a single night in January organised by a company called "Student Republic" who denied that they were encouraging binge drinking. A surprisingly unconservative *Gazette* editorial came out in favour of the "ambitious, youthful Canterbury-based independent firm", pointing out that there had been no disorder reported.

11th February 2014 to 21st March 2014

Parts of the city centre were cordoned off by an MoD bomb disposal unit for five hours on February 13th when a parcel bomb arrived at the Army recruitment centre in St. Peter's Street. This was one of a series of crude devices sent to offices across southeast England by the "New IRA". A couple of days earlier, police had been called to Burgate after reports of men seen on the roof of a shop...who turned out to be roofers! Perhaps some kind of premonition was involved here as a few weeks later a bearded man in a red jumper was seen leaping and climbing on the roofs of shops and restaurants. "Canterbury's Spiderman" was initially seen above St. Peter's Street and then later the same day on buildings in Sun Street. Evading arrest, it wasn't clear whether he was looking to break in somewhere or just "urban free climbing".

Each Saturday during this time a group of local Christians set up chairs on the corner of Best Lane and the High Street and offered to cure illnesses through prayer, claiming to be performing miracles. Claims circulated that they'd been telling people not to take vital medications and instead just to pray. "Canterbury Healing on the Streets" was part of a wider national movement. Local coordinator Simon Redman had also been running the Street Pastors project since 2011, a group of volunteers patrolling the city centre on Friday and Saturday nights with blankets, bottles of water and flip-flops, looking after (in the name of Jesus) revellers who were too drunk to function. Over at Limes Lounge on Rosemary Lane, Canterbury's only gay bar (opened in 2013), an unknown vicar had dropped in and offered to "cleanse" its customers if they came to his church. Limes had also received arson threats and owner Tony Butcher claimed that the council's threat to withdraw his license had more to do with homophobia than noise complaints. Naturally, the council vigorously denied this claim.

The rubbish collection situation became increasingly farcical when it was reported that Serco workers weren't allowed to carry bags even short distances (in case of injury), so rubbish wasn't being collected from houses to which their lorries couldn't park close enough. This affected Notley Street, Clyde Street and Alma Street. The situation even got ridiculed nationally in the pages of the *Daily Mail*. An entirely different kind of waste disposal story involved a fifty-foot headless and tailless sperm whale seen being driven on the back of a lorry along the A2 and into Canterbury one day in late February. It had died stranded between Whitstable and Sheppey, been towed to Sheerness docks and autopsied, and was being taken to the Canterbury landfill site. One astonished witness described "...*blood and guts dropping off the back and everything. The smell was unbelievable.*"

To celebrate the spring equinox, my friend Miriam and I got the Dover train to the village of Shepherdswell and then walked the nine miles or so back along the old Pilgrims' Way into Canterbury on a bright but intensely windy day. This was an extraordinary journey, both of us feeling most peculiar (possibly something we ate) as we ventured over Barham Downs and towards the Cathedral. After stopping off for a quiet moment in St. Martin's churchyard and then for some Mexican food at the recently opened Club Burrito in Butchery Lane, our pilgrimage finished perfectly with a barnstorming Lapis Lazuli concert at The Ballroom.

early 21st century

21st March to 20th April 2014

In a £7m deal, CCCU took ownership of the old prison in Longport to convert it, in part, into student accommodation. The city was now claimed to have the highest student population per capita in the country with UKC and the ever expanding Christchurch now contributing roughly equal numbers to the approximately 40,000 students in the district. CCCU had become a lot more slick and businesslike than the modest Anglican teacher training college I remembered from my student days, so I'd been surprised when it launched an MA course on "Cosmology, Myth and the Sacred", touching on a lot of material that's largely frowned on by both academia and the Church. When invited to give a guest open lecture to the students on music and mathematics, I was pleasantly surprised to see that among the members of the public in my audience were Jamie Dams and Josh Magill from the band Kairo. They were to launch their debut EP with a performance at Bramley's on Orange Street just over a week later. I arrived just in time for that on a coach back from North Wales, having spent the journey reading from a few Canterbury history books I'd recently ordered, the idea for this book having started to take shape. My experience of both the gig and the streets of the city that night were noticeably coloured in my mind by a newfound sense of historical continuity.

With this project in mind I started paying more attention to my surroundings and picking up on odd little details of city life. Cycling down Watling Street towards the Ridingate one evening, for example, I overheard an unlikely looking MC, a balding middle-aged geezer, cheerfully rapping surreal lyrics at the passing traffic: "*Here's the boss with a mouthful of emeralds / a Maltese cross and a pocket full of chemicals / the killer gorilla with the perspex hat / says I say so... and that's that*" (I was slightly disappointed when I later discovered this to be from punk poet John Cooper Clarke's "Thirty Six Hours").

Paul Fryday, a familiar old punk on Canterbury's streets, raised the alarm when the council began felling dozens of trees in the Westgate Gardens. Rejecting claims that they were diseased he argued that the council were trying to "corporatise" the gardens ("*They want straight lawns and flower beds, they don't want nature.*") He stuck posters on the remaining trees to warn of their imminent removal. The council described their actions as part of a "tree management programme" funded by a recent £1m lottery grant, claiming that there'd be more trees at the end of the project, including a community orchard. Meanwhile, the residents of Harbledown were celebrating the saving of No Man's Orchard, plans for a Park & Ride site there having been dropped.

The Campaign for Democracy in the Canterbury District launched a petition with the aim of getting 5665 registered voters to sign it so that a referendum on local governance would have to be held alongside the next general election. An example of the kind of unpopular decision-making which led to this was the council's approval of plans for 15,000 new homes across the district, including 4000 on farmland south of the city. Homelessness campaigners were angered when the *Kentish Gazette* published a sensationalist and moralising piece about a heroin addict with a violent past who'd set up camp in the bushes on St. Peter's Roundabout (fire crews had been called there twice as a result of fires he'd lit).

20th April to 13th May 2014

Despite pleas from charities to respect his vulnerable status, local news headlines continued to be dominated by the homeless man living on St. Peter's Roundabout. Fire crews had been called in five times by this stage, supposedly costing taxpayers nearly £10,000. Clearly referencing this story, CCCU student prankster Shayan Shayegani, a.k.a. "Speedo Shy", pitched a tent in the middle of the (entirely vegetation-free) Westgate mini-roundabout on Good Friday, provoking a mixture of anger and laughter. One passer by described how "*Two males dressed only in pants came out of the tent.*" He was seen there again lifting weights the following Tuesday.

With a total of 91 CCTV cameras now operating (costing more than £2m) campaign group Big Brother Watch claimed that the city was now the "most watched in the world" and called for less monitoring, more lighting and better policing. The council took a proactive approach to policing the litter situation by appointing a team of anti-litter enforcers working for contractor Kingdom, wearing stab-proof vests and empowered to give out £80 fines (half going to their corporate bosses) but with no powers to search or arrest. Not everyone was happy with this, and plans to start charging households £44.60 for their recycling bins provoked further anger. Waste collection contractors Serco were publicly shamed when it was revealed that the charity pot they'd pledged to donate to each time they missed a collection contained only £4.29, six months and 1500 missed collections later. To make amends they donated £10,000 to the homelessness charity Porchlight.

Bosses at Canterbury's Pilgrims Hospice near the London Road roundabout faced a major backlash when they announced plans to close the sixteen-bed care unit, claiming that the centre (opened by the Queen Mother in 1982) was no longer fit for purpose and that closing it would save £500,000 annually for the organisation. A Save Pilgrims Hospice Facebook group was launched and local MP Julian Brazier urged hospice trustees to find a way to keep it open.

The Piano In The Woods project I'd been facilitating reached its conclusion one night in early May when pianist Sam Bailey and "lithophonic percussionist" Toma Gouband collaborated on an extraordinary improvised recital of percussive playing on the strings of the piano (now on its back, the keys having become swollen from the damp and thus unplayable many months earlier). A PITW exhibition had opened down at the Sidney Cooper Gallery a few days earlier at which I made the acquaintance of Kevin Ayers' daughter "Lady" Rachel (living locally, and the titular subject of perhaps his best-loved song). This was all part of the biannual Sounds New Festival which also involved the Brodsky Quartet and others reworking Robert Wyatt's 2003 *Cuckooland* album at the recently opened Colyer-Fergusson music hall on the UKC campus. Later that month Lapis Lazuli launched their new album *Alien* at The Penny Theatre (known for its "low-class entertainment" in the 19th century). That night's entertainment also included a live electronica set from Koloto (a.k.a. Maria Sullivan) and my eccentric alter-ego, labcoat-clad Professor Appleblossom, introducing the band with a twenty-five minute lecture and slideshow on the "evolution and standardisation of the 'alien' archetype".

13th May to 1st June 2014

The recently appointed city centre litter wardens were already being accused of using "Gestapo tactics" to catch and fine people. Anyone disputing a fine was immediately reported to the police. Gemma Calver, a chef at Deeson's on Sun Street, was approached after dropping a cigarette butt in an alley by wardens who'd been watching her via a window reflection. When she refused to give her name and walked back to work, police officers arrived to enforce the fine. 121 tickets were issued in the first twelve days of the scheme, generating £9680. The council had only issued fourteen litter tickets in 2012, a year in which the same crew had issued over 4000 in Maidstone.

One Sunday night a Mini Cooper smashed into the RSPCA charity shop in St. Dunstan's Street. No one was hurt, but the 24-year-old driver was charged with drunk driving. When he later appeared in court it transpired that he'd got in the car to go and buy a kebab from a place only 200 yards from his home in Station Road West. Earlier that day, Kingsmead Field campaigners had staged a community rounders match on the field, this time the opposing teams being "Water Voles" and "Slow Worms", each captained by a sympathetic city councillor. Slow Worms won despite Water Voles making a dramatic comeback. More than fifty players were involved and the team names reflected the campaign's focus this year on the biodiversity of the field.

On May 31st, the day that a five-night run of a stage adaptation of the American 1950s-themed TV series *Happy Days* ended at the Marlowe, the annual World Naked Bike Ride made its second appearance in Canterbury but with strict rules this time due to complaints of inappropriate behaviour to police the year before. Those seeking only to look at naked people were told to stay away and male participants were told not to pose naked with spectators and to cover themselves if aroused. The organiser admitted that a small minority had behaved inappropriately the previous year. The nudity, he explained, was to highlight the vulnerability of cyclists on the road, not an excuse for exhibitionism.

I left Canterbury and headed to Wisconsin for a few weeks in mid-May. While there, I ordered more Canterbury history books and maps. Ideas for this book project really started to take shape during this time, including the spiral-based division of time which came to me while staring at a timeline I'd sketched out on a long strip of paper to represent over 2000 years of history. But I was left wondering how to choose the final moment, the point at which the spiral would be centred. I wanted a point in the near future of some significance but wasn't sure whether to try to instigate some kind of event or instead just to look for a date/time of some intrinsic significance in the life of the city. With this on my mind, I travelled down to the city of Madison to see my friends Syd Arthur who were touring the USA supporting John Lennon's son Sean's excellent psychedelic band The GOASTT, resulting in a pleasing collision of my Canterbury and Wisconsin friendship circles. Discussing the concept of this book with singer/guitarist Liam afterwards, in particular the dilemma of how to handle the final moment (whether to set something up or just allow things to happen), his advice was to "just let it be, Matt". Despite the proximity of John Lennon's son, the (unconscious?) Beatles connection failed to register with me for over a week.

1st June to 17th June 2014

During the rest of my stay in Wisconsin, delving into my books and maps, I began to entertain the idea of a suppressed Marianist cult in the Canterbury area. I learned that England was once known by European Christians as the "Dowry of Mary" (or "Virgin's Dowry") and that several of the early Archbishops of Canterbury dedicated much of their time to composing devotions to Mary. At the same time, I was discovering that at its medieval peak, about a third of the city's 20+ churches were dedicated to Her, including chapels above city gates. I started to remember how many times I'd walked or cycled a few miles out of the city and come to a village church of St. Mary (Patrixbourne, Chislet, Bishopsbourne, Fordwich, Nackington, Chartham, Stodmarsh, Wingham). Pictured in some of my books were figurines unearthed in St. Dunstan's of Dea Nutrix, the hybrid Romano-British breastfeeding mother goddess. If she'd had a strong local following, it seemed plausible that Augustine's missionary entourage would have stressed the only mother figure in the story they were peddling and that she (Mary) might have become the primary subject of devotion in the newly Christianised Cantwaraburgh area.

Playing with the parameters in some spiral-generating computer code I wrote to try to get a pleasing subdivision of history into manageable periods, I arrived at mid-August 2014 as my terminal point and eventually settled on the 15th, this being the festival of the Assumption of the Virgin. I discovered online that the curious, tiny shopfront "Anglican Catholic" church of St. Augustine in Best Lane still celebrated this but, having been removed from the mainstream Anglican calendar many centuries ago, it was not formally observed in the Cathedral. Although I didn't intend to be present, the Anglican Catholics were going to be holding an Assumption mass at noon on the 15th, so I opted for that moment. I'd never given much thought to the Assumption before – I'd just been looking for a moment of some astronomical or ecclesiastical significance in mid-August and it presented itself. But I later realised that the Assumption can be interpreted as the "Great Mother Goddess" character (superimposed over Jesus's mother, about whom the Bible tells us very little) being absorbed back into the creative matrix, a.k.a. "Heaven", *which she herself embodies*. The recursive, "looping" nature of this image appealed to me, reminding me of a spiral endlessly winding around a point.

After tentatively finalising this decision, I was listening to The Beatles' *Let it Be* album (1970) and suddenly realised that Liam's casual advice given in Madison linked to a lyric that includes an enigmatic reference to "Mother Mary". Paul McCartney has explained that it was a reference to his own mother (called Mary) but it can't be denied that many millions of listeners have interpreted it in a more archetypal way. I was quite happy with how this had worked out.

Meanwhile, back in Canterbury, bosses of the Pilgrims Hospice charity were reconsidering their plans to close the local "Mother Hospice" after trustees became extremely concerned about an outpouring of opposition at a recent public meeting on the matter. 500 people had turned up, three-quarters of whom then walked out in protest when trust chairman Dr. Richard Morey said it wasn't a consultative meeting but merely a presentation of the trust's views.

17th June to 29th June 2014

On the morning of June 19th, hundreds of people queued for hours for the opening of a new three-storey outlet of the largely sweatshop-sourced cheap clothing retailers Primark in the Whitefriars complex, eagerly anticipated by many in the city. A 17-year-old from Faversham was the first through the door, having queued since 8 a.m. ("*I can't believe I managed to be first. It is my new claim to fame and my mum is going to be so proud.*") Lord Mayor Ann Taylor talked to management and staff before opening the store by cutting a ribbon. In the next edition of the *Kentish Gazette*, cynical pseudonymous columnist Harry Bell wrote:

"*Their feet burning, beads of sweat dripping from their brows as the early summer sun beat down on them, they came in their hundreds. To Canterbury. Like Chaucer's characters from more than 600 years ago, they too are pilgrims. As they rounded C&H Fabrics into Rose Lane, the object of their journey rose majestically in the distance, the light shimmering from the metal and glass of the cathedral of consumerism before them. And like Chaucer's pilgrims who came to Canterbury to pray at the shrine of Thomas Becket, they too prostrate themselves at the feet of their idol: Cheap clothing sold by Primark.*"

Around the same time, the McDonalds and Burger King in St. George's Street were joined by a KFC. A Rentokil van was photographed outside it a few days later, sparking rumours of a pest infestation, but management claimed they were proactively installing fly control units.

In the wake of popular opposition, the Pilgrims Hospice organisation decided to keep their Canterbury hospice open, although it was unclear how many of the sixteen beds would be eliminated. The chief executive and chairman of trustees both resigned in the wake of the controversy. The Get Canterbury Moving campaign group was hastily reformed after months of confusion as to what exactly the city and county transport authorities were planning for the Westgate area (pedestrianisation, pavement widening, speed limits, bike routes, pedestrian crossings, weight restrictions and roundabout removal all being discussed). The Canterbury Society joined forces with the Campaign for Real Democracy in Canterbury to hold a public meeting at the Friends' Meeting House about the way decisions were made locally.

A £1m upgrade to the Westgate Gardens and the adjacent "Toddlers' Cove" play area was unveiled, involving new bridges and a new physic garden that wasn't quite ready. The play area (originally created in 1943 as part of the wartime "Holidays At Home" morale booster) ceremoniously opened and then closed immediately afterwards as it also wasn't quite ready.

Just back from Wisconsin on the 19th, badly jetlagged, I went along to an evening of music at The Ballroom: Meg Janaway, Kairo and The Hellfire Orchestra (from Deal). The audience was initially quite small due to England playing in the World Cup that night but the place eventually filled up. Earlier that day in town I'd happily run into Phil Holmes, the sax and accordion player in Lapis Lazuli, and the lovely, unassuming but brilliant local jazz pianist Frances Knight. I was feeling glad to be back and connected to this place.

29th June to 8th July 2014

On the evening of June 30th a Sainsbury's security guard chasing a shoplifter along the riverside path received a four-inch gash to his stomach when the culprit slashed him with a blade. Around the same time, well-known mentalist and illusionist Derren Brown opened a six-night run of his latest stage show *Infamous* at the Marlowe. During his stay he took some striking photos of the city's street life and posted them online for his 1.7 million Twitter followers. He also visited a terminally ill woman in Harbledown and tweeted thanks to various cafes and restaurants for an enjoyable time in the city. During his residency the 40th international conference of the Joseph Conrad Society was taking place up the hill at UKC's Keynes College. The choice of location was relevant as the great Polish-born author had lived his final years in nearby Bishopsbourne and was buried in Canterbury City Cemetery in 1924.

The next week, shoppers in the high street were treated each day to unexpected "flash mob"-style dance routines (with dancers disguised as builders, shoppers and businessmen) staged by the Protein Dance company, part of the Kent Dancing festival sponsored in part by UKC's Gulbenkian Theatre. This was billed as two weeks of dance events "in unusual places" (including museums and galleries as well as the city streets).

The future of the building adjacent to the Westgate at 1 Pound Lane finally became clear after a long period of uncertainty when two young local entrepreneurs won the tender process to transform it (formerly a city jail and police station) for public use. They had plans for an upmarket restaurant and bar. But confusion was still swirling around the various conflicting plans of the city council and Kent County Council over the future of the Westgate road system layout. Temporarily complicating the situation, the barriers at the nearby St. Dunstan's level crossing failed twice on July 1st, causing extensive diversions and queues. A couple of days later a homeless 39-year old was fighting for his life after having his head smashed on the pavement outside the nearby Falstaff Hotel while in a dispute with a younger homeless man.

On July 3rd the council's ruling executive voted to spare 75% of Kingsmead Field from development (the other 25% was to be auctioned off to developers). Protestors had mixed feelings about this partial victory, as council leader John Gilbey was not convinced that the field should be given Village Green status (so the undeveloped portion of the field could potentially be sold off in the future). Meanwhile that evening I was helping to facilitate an intimate semi-acoustic concert by Syd Arthur at a secret woodland location near town. This involved stripped back versions of songs from their second album *Sound Mirror* and was being filmed by an American crew for the documentary series *Off Main Street*. A few days later the band was flying back off to the USA to support prog-rock legends Yes on a national summer tour.

On the morning of Monday 7th July, Police and Trading Standards raided the headshops Skunkworks (Northgate) and Third Eye (St. Peter's Street), confiscating various "legal highs" on sale and taking them away for testing. Both shops had been earlier sent letters asking them to stop selling these products.

left: the third Marlowe Theatre, built on the site of its demolished predecessor and opened in 2011; *right*: street artist Stewy's stencilled portrait of Robert Wyatt which appeared in Dover Street in 2012

left: the "Crows v. Seagulls" cricket match held on Kingsmead Field in September 2013 to raise awareness of the campaign to save it; *right*: past and present Liberal Democrat city councillors supporting the Kingsmead Field campaign

left: one of countless graffiti tags by Oreo (a.k.a. Kitt Klarenberg) which appeared around the city in 2009-10 before he was caught and fined for his efforts; *right*: the labyrinth opened on the hillside below UKC's Eliot College in 2008, designed by Jeff Saward and Andrew Wiggins in traditional "four-fold" medieval style

left: prog-rock maximalists Lapis Lazuli in the garden of their bungalow on New Dover Road: Dave Brittan, Dan Lander, Adam Brodigan, Phil Holmes, Neil Sullivan, Cameron Dawson; *right*: electronic musician/producer Maria Sullivan, a.k.a. Koloto

left: Syd Arthur, outstanding in their field of psychedelic/progressive rock: Raven Bush, Joel Magill, Fred Rother, Liam Magill; *right*: chamber folk quintet Arlet: Ben Insall, Aidan Shepherd, Rosie Holden, Owen Hewson, Thom Harmsworth (like the classic 1970-71 Soft Machine lineup, three members were graduates of Simon Langton Grammar School)

left: Langtonian Afro-prog funksters Zoo For You: Josh Magill, Vinnie O'Connell, Barney Pidgeon, Andrew Prowse, Bruno Burton, Owen Hewson, Thom Harmsworth; *right*: Canterbury Sound revivalists The Boot Lagoon: Pete Edlin, Seth Deuchar, Cameron Dawson, Callum Magill

left: Orange Street Music Club, which occupied the room above the old Prince of Orange tavern (compare picture on p. 83); *right*: *Pebble For Nicholas Heiney*, sculpted by John Das Gupta and installed in the Cathedral Precincts' Kent County War Memorial Garden around 2010 (Heiney was a talented poet who took his life in 2006 at age 23, the son of BBC radio broadcaster Libby Purves)

left: The Pilgrims Hospice on the London Road, controversially threatened with closure in the spring of 2014 but saved by an overwhelming public outcry; *right*: Westgate Hall, a community space since the early 1900s, repeatedly threatened with demolition by Canterbury City Council but eventually saved by the community in 2011

8th July to 16th July 2014

Saturday July 12th saw Whitefriars host an "Our Big Gig" event, part of a national scheme seeking to improve community cohesion and inspire people to take advantage of local music-making opportunities. This was curated by Music For Change, a local charity promoting intercultural understanding through music, and included local acts Magga Tiempo and Mahalas. Meanwhile, over at the Westgate Gardens, the Rotary Club was holding its annual charity duck race on the Stour. 3500 small yellow plastic ducks were thrown into the river at Toddlers' Cove with a noisy crowd at the finish line cheering on the winners. £45,500 was raised, to be split between four local charities. The next day, the Kent Dancing festival culminated with the finale of Protein Dance's "(In)visible Dancing" show on the high street, featuring members of East Kent's Argentine tango group, Westcourt Tango. This was enthusiastically received by the audience despite heavy rain. Whitstable Choral Society performed Handel's oratorio "Israel in Egypt" in the Cathedral on the 14th. Earlier that day, comedians Sandi Toksvig and Harry Hill had received honorary doctorates from UKC at a ceremony there. The next day a five-night run of *West Side Story* began at the Marlowe.

The Save Kingsmead Field campaign soldiered on, challenging the council's decision to sell 25% of the field to developers. It seems that the report on which it had based that decision had failed to take into account the land's biodiversity or recreational value. SKF leader Sian Pettman had also been active in the Campaign for Democracy in the Canterbury District, which succeeded in getting a council committee to recommend that the authority switch from its one-party leader and executive system to a committee structure reflecting its overall political makeup. Jo Kidd, coordinator of the Canterbury Alliance for Sustainable Transport, expressed her support for the council's latest iteration of its traffic plans for the Westgate area but suggested that more could be done, with St. Dunstan's Street turned into a King's-Mile-like "iconic and vibrant entrance to our historic city". On the 10th, teachers, firefighters and other public sector workers from the PCS, Unite, GMB and FBU unions had marched through the city to protest pay freezes and changes to pensions as part of a national day of action, chanting "*No ifs, no buts, no public sector cuts!*"

Strip club The Bing applied for a renewal of its license after ten trouble-free months. The two complaints from residents' associations were dismissed by the *Gazette*, which opined that "*Canterbury is a young, vibrant and increasingly fashionable city. It is not the personal domain of those who want to turn the clock back to the 1950s.*" Turning the clock back to the Middle Ages, it was announced that a commemorative statue celebrating Chaucer and his *Canterbury Tales* was being designed for the High Street. Following the lead of local celebrity Orlando Bloom, department store boss (and High Sheriff of Kent) Hugo Fenwick paid £5000 to have his face featured on one of the pilgrims – the Merchant, of course.

Council contractors were called in when a large tree fell into the Stour, blocking the stretch between Rheims Way and the Westgate Gardens. They cleared its roots from the riverbank but left the tree in place to be removed a week or so later, declaring it stable and the area safe.

16th July to 22nd July 2014

Friday July 18th saw an anomalous weather front called a "Spanish plume" roll in over Canterbury from the south. Cycling up to the slopes of the UKC campus from Wincheap I could feel something big was coming in, periodically glancing back over my shoulder to see what looked like a bright white roller blind being pulled steadily up the sky. By the time I was below Eliot College there was a thunderhead that looked like something CGI'd into a fantasy film, with lightning flickering out from the top of it. This was weather unlike any I'd seen before. "Apocalyptic" weather conditions were said to have hit the 40th anniversary Kent Beer Festival at Merton Farm that evening although people carried on dancing to a 70s cover band called Disco Inferno. Earlier in the day, down at the Cathedral, Canterbury Scene legend Robert Wyatt had received an honorary doctorate as part of a University of Kent graduation ceremony. My friend Sophie, working for the University's events office at the time, was partly responsible for his well-being and by all accounts he seems to have enjoyed his stay in town. When asked if there was anything else that anyone could do for him, he'd poignantly joked that he wished they could bring his father back from the grave to see that his son hadn't been a complete failure!

Billy Smart's Circus (established 1946) rolled onto Kingsmead Field with dozens of lorries and caravans. This caused significant damage to wildlife habitats, flattening a meadow which had been left uncut to encourage slow worms. Campaigners were reported to have called this a "deliberate act of vandalism". They later clarified that they welcomed public events on the field but urged discretion regarding their scale, suggesting that circuses should be held elsewhere. A year earlier, a monster truck and motorcycle stunt show had left scorch marks and a sea of litter on the site. Unlikely to win over the local demographic that had attended that event, the Green Party issued their response to the council's recently published Local Plan, suggesting that Canterbury should strive to be the first car-free city in the UK, becoming a "beacon of eco-living" by 2031. The newly formed East Kent Against Fracking group held their initial public meeting at Thanington Neighbourhood Resource Centre, council leader John Gilbey (a retired geologist) still sitting on the fence with regard to this controversial issue.

After much discussion local traders voted to establish a "Business Improvement District" (BID), meaning that they would put £500K of their own money into improving the city centre for their supposed mutual interest. 708 businesses were balloted but only 381 voted (242 for, 137 against) so trader/activist Debbie Barwick vowed to appeal the decision as unethical. Meanwhile, the Wincheap Society appealed to the council and universities to help solve the problem of students leaving excessive rubbish and junk outside their rented homes at the end of term, the idea of employing "street marshals" being put forward.

As time began to hurtle towards my chosen date of 15th August I found myself spending more time in town just observing. Someone I kept seeing around the Burgate area during this time was a homeless Irish guitarist called Ziggy, always wearing a green felt hat. My friend Lucy's Wild Goose tapas bar in the Goods Shed became a favourite hangout for sipping beer, eating olives, editing my notes and surveying the scene.

22nd July to 27th July 2014

On Wednesday July 23rd much loved local punk Davee Wild collapsed and died in London, aged just 32. He'd been selflessly promoting punk gigs in Canterbury for years, mainly at The Maidens Head in Wincheap, after first becoming acquainted with the local scene at the Cardinal's Cap in Rosemary Lane as a teenager. I didn't know Davee – there were so many disjoint (as well as overlapping) music scenes in the Canterbury area. The next evening, Miriam and I were at The Ballroom to see Colonel Mustard, Boxing Octopus and a new reggae band called Hey Maggie. We were both struck by the number of unfamiliar faces, guessing that many were followers of Boxing Octopus from Herne Bay. The same night, organist and Radio 2 presenter Nigel Ogden played at St. Mary Bredin church (in Nunnery Fields, the original building in the Whitefriars area having been destroyed in the Blitz) accompanying the silent film *Nosferatu* in aid of Pilgrims Hospice.

That morning, a 51-year-old Herne Bay woman had called police to say she'd left bombs outside the Cathedral, council offices, Magistrates Court and both railway stations, that she could detonate them by mobile phone and that it was up to the police to find them. Nothing was found, the call was traced back to her and she was arrested. (A couple of weeks later, Judge Adele Williams would adjourn sentence for psychiatric and other reports to be prepared.) Meanwhile the same morning, retired social worker Jenny Bell of St. Dunstan's was shocked to see a lorry drive towards the high street from The Friars, narrowly missing someone in a wheelchair. She called the city council about this, who referred her to Kent County Council, who told her to contact Kent Police, who suggested she get in touch with the city council. That afternoon police showed up at the Dane John Gardens after reports of a large brawl. Many involved fled the scene but a 24-year-old Londoner was arrested on suspicion of affray.

Over at the Magistrates' Court, a 33-year-old was given a four-year prison sentence, having been arrested with £8000 worth of cocaine at his house in Pine Tree Avenue back in March. It was reported the same day that PC Brett Wright, who'd been caught on CCTV punching a young shoplifter in the face, would not be prosecuted. There was supposedly not enough evidence to "provide a realistic prospect of conviction", a judge at Maidstone Crown Court decided.

On the afternoon of Friday the 25th, a driver suffered shock when her car collided with a lorry on Wincheap Roundabout, trapping her inside. Parts of Pin Hill and Rheims Way were closed for half an hour while she was freed. At 11:30 that night a man was taken to Ashford Hospital after suffering a head injury when assaulted in Lower Bridge Street.

For the youth of the economically deprived Sturry Road area, the old basketball court at the Sturry Road Community Park was reopened on the 26th after having been given a facelift. Opened in 2007 on a capped former landfill site, the grounds now included a grass five-a-side football pitch, skatepark, mini BMX track, wildlife meadows, ponds and woods. A few days earlier the Marlowe Youth Theatre group had been visited by their new patron Orlando Bloom at the theatre that had replaced the one where he'd made his stage debut at age four.

27th July to 31st July 2014

It was raining heavily on the evening of Sunday 27th July while I spent a tranquil evening with friends in Martyrs' Field Road playing the ancient Chinese board game Go – a regular Sunday evening thing at this time – with Jethro Tull on the stereo. The next night, the A2 was closed for three nights for resurfacing from its nearby Wincheap junction out to the M2 motorway, while the Rod Stewart musical *Tonight's The Night* (written by comedian Ben Elton) began a six-night run at the Marlowe. On the 29th, Kent Count Council announced that Kingsmead Field was not going to be granted Village Green status. Campaigners were "disappointed but not surprised" and expressed their intention to put up football goals, create a children's play area and plant more trees on the remaining 75%. Around the same time, the city council announced that vehicles from Billy Smart's Circus *hadn't* damaged the field and that it should recover, but that the monster truck show scheduled for August was to move to the old coach park nearby.

For several months a Canterbury edition of the boardgame *Monopoly* had been in development, including a competition to award a square to a local charity. It was announced that "Old Kent Road" (the cheapest on the board) was to go to homeless charity Catching Lives, responsible for the Canterbury Open Centre in Station Road East. This came after concerns were expressed that the city might be "too posh" for an Old Kent Road. James Duff of Catching Lives told the press that "*being one of the cheaper squares will help with our message:* Monopoly *is all about money, and we work with people who tend not to have very much.*"

On the 31st two French teenage girls visiting the city were attacked in the Dane John Gardens by two local teenage girls. The pair were swiftly arrested and scheduled to appear in Canterbury Youth Court in August. Days earlier, two local 21-year-old men were jailed for carrying out several random violent attacks in the city centre the previous September. One victim had been attacked because he "looked gay" and another had lost his front teeth. Judge Simon James told the pair that "*Canterbury is a small city with a justified international reputation, attracting people from all over the world to study and enjoy its cosmopolitan atmosphere and its many bars and pubs... It seems to me that people are entitled to walk the streets in safety without being attacked... Conduct like yours is an unacceptable blight on our city.*"

Around this time I began testing a camcorder I'd bought for the final stages of this project, intending to discreetly record the life of the city as noon on August 15th approached. Spending more time wandering Canterbury's streets, I ran into various familiar faces: half-Catalan Xiana from the Wincheap Go scene walking along Burgate Lane on her way to the Job Centre; my musical collaborator Tom Holden (bass player in the now defunct "disco folk" band Famous James and the Monsters) with his punk brother Ed and Dan Lander from Lapis Lazuli outside The Cherry Tree pub on White Horse Lane. But the more I attempted to dig for undercurrents of subversion and weirdness, the more it felt that an atmosphere of *total normality* had settled on the city, quite disconcertingly. Walking into town with Miriam a week earlier she'd disparagingly remarked that "nothing ever happens in Canterbury". This had slightly troubled me at the time but now seemed to be becoming my experienced reality.

31st July to 3rd August 2014

Senegalese musician Nuru Kane, a truly captivating performer, played with his band at the Gulbenkian Cafe on the UKC campus on Thursday the 31st. It was such a warm evening that the cafe doors and windows had been opened, effectively making it a free event. I sat on the grass outside and enjoyed it from the open air, being continually surprised by the disinterested passersby who didn't even stop for a second or turn their heads to look at this amazing, full-power African music machine a few feet away from them. And I was equally surprised by how almost no one I knew had bothered to come out for this. I'd been sitting working in the Gulbenkian many evenings of late, sipping ale and peoplewatching over my laptop screen. There were a lot of Dutch around at the time but they weren't young language students (who flooded the campus each summer). These were middle-aged, sitting around playing cards, chatting and drinking, possibly over for some kind of conference – I never found out the story.

Lounge on the Farm festival happened at the weekend. It had been reduced to just a relaxed one-day event, which a lot of people seemed to think was an improvement on previous years. The organisers stated that it was to be the last LOTF at Merton Farm but it turned out to be the last one anywhere. The lineup wasn't particularly stellar: Peter Hook and the Light, Fun Lovin' Criminals, Courtney Pine, Dub Pistols. The night before, not too far away at Vernon Holme (the former home of 19th century painter Thomas Sidney Cooper), the local Rotary Club had organised a charity music evening of trad jazz and R&B, featuring the Noel McCalla Acoustic Trio. This raised £1000 for Save the Children, particularly for the children of Gaza who were being bombed *again* in the middle of Israel's seven-week assault on the Strip.

I'd recently had "persistent contrails" (or "chemtrails", depending on who you asked) pointed out to me by my dear friend Rosemary of Harbledown. Around this time I was seeing many blue skies over Canterbury being crisscrossed with these trails which gradually fanned out into a milky haze. I'd looked at both the alarmist websites and the sceptical ones and didn't know what to make of the issue, but it added to the mild feeling of background dread I was starting to feel about events to come as the now significant-seeming date 15th August 2014 rushed towards me.

On the evening of August 2nd I went to meet my friend Sarah F (from 90s Whitstable days) and her coworkers from Eurostar in Ashford outside Canterbury West station, bumping into Miriam there, on her way to Austria to perform with her Little Bulb Theatre company, via the Battersea Arts Centre. Sarah's friend Mike Nistor had written a short comical play called *My Day Off* which was being debuted at the Marlowe Studio theatre as part of an event called *Modern Heresies*, so we went along and enjoyed that along with a couple of other short plays. As we left, a great mass of people began flooding out of the main auditorium, many in paper sailor's hats. I later deduced that it was the last night of the Rod Stewart musical, the hats no doubt distributed during "We Are Sailing". I dropped my bike lock key and ended up groping around for it in the dark for a few minutes while Sarah and her friend Matt waited under the Marlowe memorial statue which had caused such scandal back in Victorian times. A few of us then ended up back at her house in St. Dunstan's, drinking wine and chatting until late.

3rd August to 5th August 2014

I helped to facilitate another intimate, free woodland concert on some friends' land near town on Sunday the 3rd. This featuring the trio Kairo, also with short sets from singer-songwriter Meg Janaway and traditional folksinger Sarah Y (joined for a couple of songs by our mutual friend Claire). It was a happy occasion, with friends Kim and Adam (he the drummer in Lapis Lazuli and Delta Sleep) staying over in the woods with me that night. A couple of days later they invited me out to their rural home in Barham, letting me know that there was a lift going from the Lapis Lazuli HQ bungalow. On my way there, cycling down the Old Dover Road, I encountered Nick Dent. He'd been one of the first people to speak to me when I was a student at UKC (a committed Marxist all the time I've known him, he was trying to recruit me for the Socialist Workers Student Society). Nick had been working in mental health for some years and had just been to a training workshop about dealing with people who hear malevolent voices in their heads.

On the Save Kingsmead Field Facebook page, campaign leader Sian Pettman had updated that day as follows: "*The campaign is now in its third year. We're having a campaign break from 1-25 August to build up for the next stage of the campaign in the autumn. The Save Kingsmead Field Group has been set up to campaign to save the last significant piece of public open space in the Kingsmead area of Canterbury from being sold off for housing development by Canterbury City Council.*"

On August 4th the Canterbury Old Stagers began a six-night run of the dark comedy *Arsenic and Old Lace* at the Gulbenkian Theatre. Now claiming to be the world's oldest continually operating amateur dramatics group, they'd been performing plays exclusively in the city since the 1840s when they'd been affiliated with the "I Zingari" nomadic cricket team. At 11 that morning a crowd had gathered at the Buttermarket for a service to mark the centenary of the start of WWI. Relatives of the fallen placed small wooden crosses below the memorial and at 11 p.m. the lights of the Cathedral and other public buildings were dimmed as part of a "Lights Out" national commemoration to mark the hour that Britain declared war on Germany. The Sidney Cooper Gallery was appealing for local WWI photos for a possible exhibition later in the year.

The next afternoon another senseless death occurred when 36-year-old software developer Christopher Atkins of Salisbury Road was killed when hit by a train at 12:45 p.m. at the Tonford Lane level crossing in Thanington. He was pronounced dead at the scene, the circumstances not considered suspicious.

Around this time the city council published results of a survey on the services they provided. Only 4% of people polled in Canterbury said that they highly valued the roles of Lord Mayor and Sheriff of Canterbury but the council said these were "part of the fabric of the city" and so had no plans to scrap them. But 85% declared themselves "satisfied or very satisfied with living in the district". Only 4% "strongly agreed" that the council understood its customers and 22% didn't think they could be trusted. The demographics of this survey were rather skewed, though, with only 3% who responded to it being aged 18-25 and 37% being over 65.

With our espisodes down to less than two days each, we're now timing to the nearest minute.

19:51 on 5th August 2014 to 18:59 on 7th August 2014

I spent the evening beside a fire near Barham in good company, then left on a drizzly Wednesday morning. Back in Canterbury I cycled through the Dane John Gardens to witness the sad spectacle of a "Family Fun" day completely washed out by the rain – quite a few gazebos but almost no one in attendance other than a few stewards. A WWI commemoration gazebo had been set up next to the blatantly phallic Boer War memorial.

With just over a week to go, I sharpened my focus and began to spend even more time in town observing. With my back to the former medieval pilgrims' Chequer of Hope Inn on the corner of Mercery Lane (now a sweetshop), I watched a dad in a "BROOKLYN HUSTLER" top wandering idly around the open space that was once St. Mary Bredman church, talking on his mobile and gesticulating while his two little boys ran around playing noisily, the assembled pigeons unfazed. Damp people. Foot-long hot dogs. Body piercings. Surprisingly few tourists. No buskers. The smell of slightly burned waffles coming from La Trappiste on Sun Street as the sun came out. A new bike shop called "BikeTart" near the 17th century "Crooked House". "REVALUTION" chalked on a wall of St. Mary's Northgate. Community archaeology notices near the Abbot's Mill Project on St. Radigund's Street (still not much visibly going on there). Ubiquitous and enigmatic street character Angus walking past with a fake gold chain around his tweed trilby, gold mirror shades, scraggly beard, guitar on his back as usual (I'd yet to see him actually playing it). Several older men in matching suits and ties ambling by looking grim (funeral attendees?). A woman on a mobility scooter whizzing past the main gate of the Westgate Gardens loaded up with pots of lilies. A dark-skinned youth suddenly appearing from nowhere and kicking a seemingly abandoned football in the Gardens, birds then ascending en masse. Ducks on the Stour. Two young Turkish men deep in conversation outside Westgate Barbers, clutching e-cigarettes. Four balding taxi drivers earnestly chatting outside Canterbury West.

That night the pedestrian tunnel between Canterbury West station and Beverley Meadow was adorned with large, angry-looking purple graffiti: "TEEN HORNINESS IS NOT A CRIME" (circled "A") and "Love yourself, arsehole!" Clearly from the same spray can, the window of Barretts car showroom near the Westgate (specialising in Jaguars, etc.) now announced "rich cunts here". The Army recruitment centre was also hit ("FUCK THE ARMY").

The next day, in the carpark next to The Causeway, I spotted a red VW campervan with a ladder and emergency light on its roof, looking like some kind of life-sized toytown fire brigade vehicle. Going over to take a closer look I found it to be in immaculate condition, with German number plates and *Order of the Garter insignia* on its doors. I never did find out what *that* was about.

A "be nice" sticker on a lamppost. A "FNORD" sticker (one of many seen around the city) on a bollard on The Causeway. Tattooed geezers, a severely intoxicated couple, trendy young men with slicked-back hair, skateboarders. "Elvis Burgers" at The Lady Luck. "Equivalenza", a new cheap perfume outlet next to the old Friars of the Sack house. "I'M THE TRUTH" on a young man's red baseball cap. Muffled drum'n'bass from a distant car.

18:59 on 7th August to 8:33 on 9th August 2014

On Friday the 8th, planning application CA//14/01610/FUL was registered for a conversion of part of the Westgate complex into a restaurant/cafe/drinking establishment. This was to involve work being done on a listed building and some new signage. On the same day, Westgate Hall's new roof was completed, paid for jointly by a grant from the Department for Communities and Local Government and an investment from Curzon Cinemas. It introduced several bright red windcatcher fins to Canterbury's skyline. The Westgate Hall Trust proclaimed online that "*their rounded cheerful shapes bring a smile... and Canterbury can't miss them... Suddenly, saving the Hall is no longer a dream or a plan. It's real. And the potential of the Hall's future is becoming tangible. A real community achievement.*" Over at the Beaney a collection of sixty photographic portraits went on show, shortlisted for the Taylor Wessing Photographic Portrait Prize organised by the National Portrait Gallery.

That afternoon I had to get a bus to Broomfield to see an osteopath so spent a little while sitting on the pavement by the bus station eating some of the previous night's leftovers from a Tupperware box, watching the Park & Ride queue. Lots of fashion victims, many overweight people, a lot of tattoos. But all very peaceful, I noted contentedly. A suited ticket inspector boarded my bus near Tourtel Road, the first I'd seen in years. He was very jovial, despite sweating profusely from the heat and humidity, chatting at the back with a passenger about his holiday in Scotland, his job and zero-hour contracts. From the window I noticed a sign for a "student village" off the Sturry Road – I hadn't realised that there was such a thing.

When I got back to the station, bus drivers in fluorescent waistcoats were smoking and eat packed lunches. I noted an audacious new graffiti tag up near the roof of the Odeon cinema – "ER LONG" – seemingly executed from a very awkward angle. In Whitefriars: an awkward kid on rollerskates; a strange inflatable gazebo promoting "motorline.co.uk" beside a red car emblazoned with "GO FUN YOURSELF"; a proliferation of Batman T-shirts like it was 1989. A new tagger had declared him/herself "all city 2014" on a wall in St. Margaret's Street. Up in the Veg Box Cafe a woman in a "Self-initiated" T-shirt and with Celtic-style tattoos chatted with the waitress at the till for a while. Downstairs in Canterbury Wholefoods I talked with employees Holly and Phil (he of Lapis Lazuli) about muscles, fascia and a physiotherapy technique called Rolfing™.

In the Westgate Gardens: sunlight sparkling on the Stour; a loud older bloke with a strong estuary accent swearing profusely into his mobile; an oriental family sitting on the grass nearby picking at takeout food; a youngish woman with white fluffy dog, multiple piercings and ivy tattooed on her back playing percussive guitar and singing passionate bluesy songs on the riverbank; Angus again (known to some as the Wizard of Westgate Gardens) with some street drinkers by the war memorial; a youth gymnastically hanging off a bridge retrieving a football from the river with his feet. On a patch of wasteground near Canterbury West, two women picking blackberries. I ran into Miriam's ex-boyfriend Piers and we talked about his progress on the shakuhachi (traditional Japanese bamboo flute) and my intensifying commitment to this strange project.

left: the mysterious German-registered "Order of the Garter" camper van seen in Millers Field Car Park beside The Causeway; *right*: Robert Wyatt receiving an honorary doctorate from UKC at a ceremony in the Cathedral on 18th July 2014

left: the extraordinary "Spanish Plume" weather formation which blew over the city a few hours after Robert Wyatt got his diploma; *upper-right*: thousands of small yellow plastic ducks being dumped into the River Stour at the start of the Rotary Club's annual charity duck race; *lower-right*: several unions joining forces, marching against the Conservative government's austerity programme

left: well-loved local punk Davee Wild who died in July, later to become the subject of annual memorial events; *right*: one of countless "FNORD" stickers which appeared around the city in the years leading up to 2014, presumably inspired by *The Illuminatus! Trilogy*

left: Local reggae band Hey Maggie: James Ross, Josh Parnell, Matthew Cox, Matthew Webb, Mark Camateras, Jules Madjar; *right*: homelessness charity Catching Lives in Station Road East celebrate their having been awarded the cheapest square on the new Canterbury edition of the board game *Monopoly*

left: the Sidney Cooper Gallery on St. Peter's Street, the property given to the city in 1868 to be an art school by the celebrated Victorian artist who grew up in poverty in an adjacent house (Rupert Bear creator Mary Tourtel studied there); *right*: Protein Dance's *(In)visible Dancing*, an unannounced "flash mob"-style performance in St. George's Street

the Odeon Cinema on St. George's Place (originally the Regal Cinema, opened in 1933) with the audacious but not-particularly-well-executed "ER LONG" graffiti tag visible down the far-right edge, "OTT" having left his/her mark at the top of the building some years earlier

8:33 on 9th August to 14:29 on 10th August 2014

A charity walk set off from the Umbrella Centre on St. Peter's Road at 10 a.m. on the 9th, completing a full circuit of the city "as far as ASDA" to raise money to support people with mental illness. In the afternoon, the Canterbury & District Model Engineering Society kicked off their "Trains and Traction" weekend in Sturry. Meanwhile in Sussex, Canterbury City FC had a disappointing start to their season, losing 2-0 to Crowborough in a rather patchy game. I had to travel to Sussex that afternoon, taking a train from Canterbury West to Lewes via Ashford. Ahead of me on the Beverley Meadow pathway en route to the station: a kid with long curly hair on sprung stilts; a gaggle of Italian language students carrying bags advertising that they'd just been to Primark; a *very* young family (two kids seemingly the results of teenage pregnancies, the father in an X-Men T-shirt); Polish men smoking cigarettes and drinking coffee. Four middle-aged cyclists on expensive-looking bikes and clad in lycra rolled into the station. On the platform, three or four very loud blokes emerged from the men's toilets together, cigarettes behind their ears, shouting amongst themselves (cocaine?).

As the train headed out through Chartham and Chilham I noticed that I was drifting towards a more detached perspective – Canterbury seemed smaller, more provincial and more mundane than it did when I started this writing project. I was becoming under- rather than overwhelmed by its history. To my surprise, I could even feel tentative "time to move on" feelings surfacing. And the next few days of my life – these feelings included – were going to become part of the narrative I was constructing, unavoidably. It was starting to sink in.

At 7 that evening a silver hatchback MG somehow flipped upside down on Beaconsfield Road but miraculously no one was seriously hurt. The driver suffered minor injuries and was taken to hospital. Bizarrely, no other vehicles were involved. Around that time, a charity sleepout at the rugby ground on Merton Lane was getting started. During the night, more than a hundred people in tents and cardboard shelters were drenched by the tail-end of Hurricane Bertha. The event raised around £2000 for the homeless, split between the charities Catching Lives and Porchlight. Seven-year-old Rosie-Lillie Smith won a competition for the best decorated cardboard shelter. The organiser explained: "*It was never meant to replicate homelessness, but I think quite a few people experienced an uncomfortable night... Next year I think I would like to throw out a more realistic challenge to local business and community leaders to sleep out in the city, but without tents.*"

The next day, the Regimental Association of the Queen's Own Buffs held their annual reunion, parading down Burgate to the Cathedral where a service of remembrance was held. The Model Engineering Society cancelled the remainder of their weekend when heavy rains hit at lunchtime. Meanwhile on the banks of the Stour, a few miles upstream at Chartham, local artist Tom Langley and friends delayed the start of their second annual "rubbish and recycled raft regatta" by a couple of hours because of the rain. Everyone got soaked by a fifteen-minute downpour after setting off towards Canterbury but the sun and wind then came in force to produce "a really dramatic nice time," according to Tom.

14:29 on 10th August to 14:21 on 11th August 2014

The rubbish raft regatta (this year rebranded as the "Crud Derby") was powered down the Stour by the remnants of Hurricane Bertha on the afternoon of Sunday the 10th. Because of all the rain, the river was running deeper than the previous year, making navigation a lot easier. Also, there were no fishermen or river tour boats to contend with. Ed C got a smile and a wave from a couple from whose riverside garden he cheekily harvested some plums. The unfortunate Jack took two hours longer than everyone else to reach the party at Westgate Gardens where some children had made a bunting-based finishing line and cheered the rafts as they crossed it. Despite emphasising that it wasn't a race, derby instigator and "social sculptor" Tom Langley was first across the line. Professional wild food forager Ed B's raft – a palette with a year's worth of empty plastic water bottles attached – was deemed the most innovative design. Unlike the multiple capsizings seen in 2013 there were no major disasters this year. I'd cycled alongside the first regatta with a microphone and busking amp, taking on the role of an absurdist sports commentator, but being away visiting friends in Sussex I unfortunately had to miss out on this one.

10 a.m. the next day up at the Gulbenkian, the week-long "Play Factory" workshop began, challenging participants (11-19 years old) to create a play in a week from scratch. A few hours later, folksinger and pilgrimage revivalist Will Parsons and his dog Holly set off from a cottage in Elham along footpaths to Canterbury, seeking a new route. In previous years Will had walked and sung as one-third of Ed, Will and Ginger.

refectory kitchen
@refectorykitch
⚙ 👤 Follow

Live music tonight with 'quiet americans' .

11:42 am - 10 Aug 2014

Canterbury Freegle
@FreegleCantbury
⚙ 👤 Follow

WANTED: barometer (Canterbury)
go.frgl.it/XD0Tg

3:43 am - 11 Aug 2014

(Tweets are timestamped with the time in San Francisco.)

Canterbury Cathedral
@No1Cathedral
⚙ Following

Many thanks to @countrylivinguk for the great stained glass article.
goo.gl/x5xlav

1:33 am - 11 Aug 2014

14:21 on 11th August to 9:22 on 12th August 2014

Around 5:30 p.m. I stepped off a train at Canterbury West. On the railings below the chestnut trees on Station Road West were tied two small bunches of dried flowers and on the ground beneath lay a small cardboard note (torn off?) memorialising a child who'd died in the bombing of Gaza in July (as had his parents and three siblings). Around this time the Shepway Singers from Hythe were singing Evensong in the Cathedral. People were huddling under awnings and trees to escape the rain. I walked across the newish footbridge over the river just downstream of the Westgate, decided to cut through Pound Lane Car Park, then somehow got a bit lost and happily stumbled into a secret-feeling urban garden which included nature-based stencil art, flowers planted in old, wall-mounted suitcases, fruit trees planted in rough wooden crates and a battered old wooden gate. A sign explained that this was the "Open Borders Garden" and that the gate used to be an orchard gate, now redundant because of the car park that replaced the orchard. Passing an entourage of foreigners with assorted umbrellas walking down Blackfriars Street, I headed to Ossie's Fish Bar on The Borough for a small bag of chips and a gherkin ("Racism Not Tolerated Here"). While I waited, fork lighting appeared directly behind the Cathedral, followed by a clap of thunder.

Walking down Castle Street, the rain having let up, I saw an approaching figure with dog, staff, rucksack and broad-brimmed hat, dressed in greens and browns. It was wandering folksinger Will Parsons, having just walked the ten miles from Elham. He and his dog Holly had entered the city near Simon Langton Boys' School. He offered me a sprig of marjoram he'd picked on the way. "*To be honest, Canterbury... it's always a slight disappointment, you know. Getting here's great, it's like striving for this ideal. You reach it...*" This led me to mention Hilaire Belloc's *The Old Road* (1904), a book he said he'd been meaning to read. He told me about a charitable trust he and his friend Guy were setting up to revive and secularise pilgrimage. They'd already got radical biologist Rupert Sheldrake as a patron. We ended up having a lengthy, deep conversation about the history, significance and future of pilgrimage.*

Continuing along St. Mary's Street, then turning down Stour Street, I saw a Dutch family struggling with the tow-hitch on their trailer. Some young men were unloading furniture from a van – I didn't realise that these were my musician friend Tom Holden's new housemates, moving into the Greyfriars estate. Wandering down St. Peter's Street I saw an open mic night being set up in The Cricketers and someone called Matt Brookes chalked up to be playing at The Black Griffin.

* listen here: tinyurl.com/y7tt5pg6

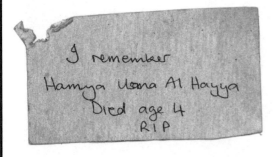

9:22 on 12th August to 00:31 on 13th August 2014

Walking to Rough Common via the UKC campus in brisk winds around lunchtime, I picked and ate my first elderberries of the year. Blackberries were plentiful by now. The preparation of the revised editions of my *Secrets of Creation* trilogy of maths books was on my mind as I noticed a digger working on something near the latest extension of the Cornwallis Building (on the exact spot where I'd shared a Portakabin office as a PhD student around '93 due to a dearth of indoor offices). Minutes later, I spotted a planning notice on a Park Wood Road lamppost for the new School of Business and Mathematics. Further down, an Eastern European man in a silver Renault, wearing a yellow fluorescent waistcoat, loose banknotes visible on his passenger seat, asked in extremely broken English if I knew anywhere he could sleep. I couldn't figure out if he meant a hotel or just somewhere he could park up and have a nap – frustratingly, he couldn't understand anything I said.

Turning down a section of the "Crab and Winkle" (Canterbury-Whitstable) cyclepath, I came upon an Orthodox Jewish gathering, something that I'd been aware was happening during each of the last few summers, although I hadn't previously known where it was based. Little boys with skullcaps and sidelocks were riding bikes and playing on a bouncy castle off in the distance while men in traditional dress stood around in quiet conversation. I noticed a wire stretched about eight feet above the lane between two posts and wondered if it could be part of an *eruv* (a ritualised enclosure sanctioned by Rabbinical law, inside which certain impractical Sabbath regulations can be bypassed). I ended up chatting with a couple walking past who'd also noticed it and who pointed out a second one further back. They pointed out the danger to "basketball players... what about bicycles, horses...?" I attempted a feeble joke about Maasai tribesmen on penny-farthings.

On my way back from Rough Common I decided to walk the labyrinth below Eliot College. Its Portland stone was a bit more worn in now, I noticed, with some lichens and mosses having taken up residence. As I sat in the middle reading about Jung's analysis of physicist Wolfgang Pauli's dreams and drinking a bottle of ale, two middle-aged women wound their way in and joined me. We talked about the soullessness of the University and about ley lines. They described themselves as "semi-local", out for an evening walk, one of them explaining that she was trying to mentally process actor Robin Williams' recent suicide.

Earlier in the day, a new ward map had been published by the council (still needing Parliamentary ratification). It involved 39 seats in 21 wards rather than the existing 50 seats in 24 wards. Council leader John Gilbey argued this would save taxpayers money and not affect democratic representation.

00:31 to 12:35 on 13th August 2014

In a state of mild anxiety I awoke from confused dreams (many comings and goings, a monastic institution, Dane John Gardens, a trip to Old Trafford). This was easily washed away with a mug of tea while I attempted to repair an old pair of headphones. Beautiful dappled light was streaming through my windows as I ate wild blackberries for breakfast. Listening to the local radio morning news for a change, I caught a story about a damning report on the K&C Hospital.

In response to four cases of tuberculosis having shown up in the local homeless population, specialist nurses from the Kent Community Health Trust were setting up that morning with a mobile X-ray and blood testing unit at Catching Lives' Open Centre (opposite Canterbury East station) to screen more than seventy homeless and vulnerable people for the disease.

Up at UKC, I passed a solitary Orthodox Jew walking across the rose garden behind Darwin College. The Templeman Library wasn't quite open yet so I sat outside peoplewatching: groups of Chinese students; female admin staff clutching takeout coffees; an interesting looking woman intently writing in a little notebook; noisy works from the West Wing; postgrads at picnic tables; a young man in a sharp grey suit eating an apple as he walked by; an aging goth clad in black consulting his iPod as he walked in the opposite direction.

From up on the fourth floor of the library it looked like something was on fire down the hill near the Archbishop's School. But it was all smoke, no flames visible, so probably just a bonfire I concluded. I spent a while making tiny typographical adjustments to the new versions of my maths books, listening online to the incomparable DJ Stashu on New York's freeform WFMU through my newly repaired headphones (I couldn't face any more local radio). After that, I prepared some graphics for a talk I'd been asked to give at my local friends Vicky and John's imminent wedding near Deal (on the challenging assigned topic "maths, love and marriage").

Willows Secret Kitch
@willowscoffee

☼   ⬩+ Follow

Introducing our Ugandan Coffee - Uganda Kinone. Huge body with dark chocolate. Tastes amazing!

3:19 a.m. - 13 Aug 2014

KentOnline
@Kent_Online

☼   ⬩+ Follow

How a woman taunted police with threats to blow up Canterbury Cathedral @No1Cathedral bit.ly/1orVuIS

RETWEET
1

10:00 p.m. - 12 Aug 2014

Kentish Gazette
@KentishGazette

☼   ⬩+ Follow

K&C Hospital branded "inadequate" in damning report as serious fears are raised about safety

kentonline.co.uk/kent/news/damn...

2:13 a.m. - 13 Aug 2014

12:35 to 22:13 on 13th August 2014

After finishing work at the library I cycled down into town. Not many people in Beverley Meadow. Overdressed, I was seriously overheating as I crossed The Causeway. An oriental woman was showing off some dance moves to her boyfriend. A kid was fishing in the Stour in the Millers Garden, a young mum taking a closeup photo of an infant in a pushchair nearby. I stopped to look at some King's School art exhibited behind a railing outside the old Blackfriars refectory while a man up on a ladder carried out some masonry repairs on the side of the building.

Down Sun Street there were swarms of tourists as usual but also four people with rucksacks, one with a walking stick, broad-brimmed hat and Ordnance Survey map. I imagined them to be pilgrims of some sort who'd just walked in but then saw them a few minutes later walking purposefully in the High Street so quite possibly they were setting off for somewhere else, starting from Christchurch Gate – I didn't think to ask. A couple of young buskers in the Buttermarket were singing Steve Miller's "The Joker" (1973).

More buskers in the High Street. A "Wake Up" sticker on a drainpipe in White Horse Lane. A few people at the picnic tables outside The Cherry Tree eating and drinking as usual. I headed up to the Veg Box Cafe for a pot of Earl Grey and more book work. Past the Belly Bar in Canterbury Lane where two girls sat waiting to get pierced, a man taking off his motorcycle helmet while talking to an employee. A new Burgate Bicycle Company was set to open on Saturday. A new busker in the Buttermarket played electric slide. I stopped in the Cathedral gate to check the noticeboard. That evening's Evensong was to be sung by a visiting choir called Angeli Roffensis.

I was surprised to see that an Anthony Gormley exhibition was on at the Lilford Gallery (also works by Billy Childish and Tracy Emin upstairs). In the middle of St. Radigund's Street I found a tiny souvenir double decker bus ("Best of British"), clearly run over, but standing up (a wheel missing, crumpled). Looking in Duck Lane Car Park for evidence of St. Radigund's Baths, I didn't find any but did then find the Judas tree beside the Stour, planted by the Canterbury Society to commemorate the Viking siege of 1011 and subsequent martyrdom of Archbishop Alphege.

Back over The Causeway. A tourist was filming ducks on the Stour. Walking up Station Road West behind a cluster of Italians, a child repeated the phrases "thank you very much, you're welcome". In the Goods Shed I found Phil Self from Deal-based band Cocos Lovers. He'd been out trying (and failing) to buy a suit for a coming wedding. Back on Beverley Meadow, a group of kids were playing football, an adult shouting instructions at them in Spanish.

Lilford Gallery
@LilfordGallery
⚙ ➕ Follow

Don't forget to come along to the gallery to see our show of Sir Antony Gormley Original Prints. All are... fb.me/3kH8q6WBU

RETWEET
1

7:40 a.m. - 13 Aug 2014

The Ballroom
@TheBallroom15
⚙ ➕ Follow

DON'T FORGET TO BOOK YOUR OPEN MIC SLOTS FOR TONIGHT!

10:33 a.m. - 13 Aug 2014

22:13 on 13th August to 05:53 on 14th August 2014

Back at home I lost myself in the FM radio band for a while, listening to minimal electronica and news from Gaza (a 72-hour ceasefire had just ended), drank green tea, made lists and tidied up.

I contemplated how I was going to spend my all-important Friday morning. Starting from the Cathedral's Christchurch gate? Walking a chain of current and former sites dedicated to St. Mary? I felt that the Abbey should be somehow involved, and St. Martin's church, for reasons of deep historical resonance. Or should I just follow my feet? I remembered that my scientific calculator had a random number generating key. The idea of using that to randomly navigate the city streets seemed appealing.

After testing my microphone/headphone and hidden camcorder setup, I tried to sleep but was kept awake for ages by a moth fluttering around me, then a mosquito buzzing around my ear driving me mad. I'd set my alarm for 5:53 a.m. to record my dreams for this page so woke up at that time with a mosquito bite on my right arm and fragments of dream mixed with fleeting threshold-of-sleep thoughts. I found myself thinking about broad-minded Americans who'd never left the USA and of sending my Belgian friend Kris in Ireland a vinyl copy of the new Syd Arthur album *Sound Mirror* for his belated birthday. I dredged up a vague recollection of a dream involving a sort of composite of my friends Sarah and Vicky who was being very helpful. It was the moment that the end of this book was to document and she was there helping to file or compile something. There was also a child who had to wear a stigmatising T-shirt with something written on it as a kind of punishment for arriving too late somewhere (functioning like a dunce's cap). I'd set up some kind of legal trust and been accused of something. All of these fragmentary impressions mysteriously tied together in a strange kind of way.

Justin Welby ن
@JustinWelby
Follow

Ending visit to Australia in marvellous Melbourne. Contrast horrors in Iraq. Pray for wise, generous not just reactive Government actions.

RETWEETS 45
LIKES 51

2:29 p.m. - 13 Aug 2014

5:53 to 12:00 on 14th August 2014

After a bit more sleep I emerged into the morning beneath a waning gibbous moon and puffy clouds interlaced with suspect streaks. It felt like it was going to be a good day, meteorologically and otherwise. After reading a bit of a Richard Feynman book over tea and breakfast, I tested my rather basic camcorder-hidden-in-cardboard-box-under-my-arm setup. Washing up around 9:25 I could hear police sirens from the Tyler Hill Road. By the time I'd packed some lunch and headed towards the UKC campus the sky was clouding over and the temperature dropping. A single magpie and a green woodpecker.

A female oriental student carrying bagels passed me in the Darwin rose garden, then two chunky tattooed blokes speaking Flemish (one in a red "Fitnesscentrum Menen" T-shirt) who really didn't look like language students or academics. The whole area around the maths building had become a building site, fenced off and with diggers to prepare the way for the new Cornwallis East extension.

In the Gulbenkian Cafe a gaggle of supervised kids were sitting around – probably part of the "Play Factory" summer theatre activity programme. I got a watery Americano to take down to the end of the recently constructed Colyer-Fergusson music annex, past a sleepy looking young woman sprawled on a sofa with her MacBook, to find a woman with a laptop talking on her smartphone in Chinese or Korean in my favoured seating area. I got on with some online work while charging my camcorder.

A group of about ten men (seemingly IT staff) came and sat across from me with coffees, talking loudly among themselves about USB devices, some kind of hack involving a particular TV tuner, firewalls, flights to North East Canada and something about WWII aircraft. A young man with red-dyed/styled hair, subtle eye makeup, gothy T-shirt and skinny jeans sat uncomfortably on the edge of the group, listening.

At about 11:40 I cycled down to St. Martin's Church for midday prayers, something I'd been meaning to do for a while. Although not a believer, I recognised the significance of this being the church with the longest unbroken use for Christian worship in the English-speaking world. Noticing fallen yew berries and confetti on the ground where I locked my bike, I went in and took a pew. An elderly Indian lady in front of me wearing a shawl said something about young people being "more efficient" and asked me to switch her phone off for her.

Kentish Gazette @KentishGazette ✿ **Follow**

THIS WEEK: Inspectors say city hospital is "inadequate" - we've got full analysis and reaction to damning CQC report

RETWEET
1

12:57 a.m. - 14 Aug 2014

Lady Luck @TheLadyLuckBar ✿ **Follow**

Live music tonight @TheLadyLuckBar thurs 14th Aug @MilkTeeth1 @theautumnravine + Muskets + Eat Me. free entry

RETWEETS LIKE
3   1

3:01 a.m. - 14 Aug 2014

12:00 to 16:53 on 14th August 2014

I found myself in the midst of a simple prayer service led by Canon Noelle West, with a handful of other people including a woman from the west coast of Canada and a couple from Tokyo. Prayers were said for various parishioners in distress, particular streets and buildings in the Parish of St. Martin's and St. Paul's Without the Walls, the situations in Iraq and Gaza, a couple of wedding anniversaries, the Parish summer fayre... I found it all surprisingly *detailed*. I joined in with the readings but slipped away at the end to avoid awkward conversation. On my way back down Longport I paused to ponder the curious insignia atop the front of the old courthouse (the *fasces* crossed with an ancient Greek "liberty cap" on a pole) and peered over the railings into the Abbey grounds where a couple were playing frisbee. I then decided to head for the silent study room in the Beaney Institute to organise my now voluminous notes.

Juliet, a recent UKC philosophy/anthropology graduate, stopped to look at a reproduction medieval map of Oxford on the kitchen wall of the house she was visiting in Best Lane. It included striking images of a dodo, a griffin and a mandrake root, the latter pictured as a shrieking humanoid being pulled from the ground of a botanic garden. Stepping out into the drizzle with her new Bulgarian rapper/producer/psychonaut friend Yanik, she set off towards the High Street, texting as she walked. At this point I noticed her passing from the window of the Beaney, initially mistaking Yanik for her ex-boyfriend Jacob (of similar height and complexion). I'd known Juliet vaguely via the UKC Psychedelics Society three years earlier but we'd never properly connected apart from the brief exchange outside the Millers Arms during that crazy storm the previous October. She'd recently moved to Ramsgate, Canterbury rents being prohibitive, and was expecting to host a group of old UKC friends on the 17th when the Japanese psychedelic/noise band Acid Mothers Temple played at Ramsgate Music Hall. This had brought her back to Canterbury to collect some bedding stored in the cellar of Sam, Jordan and Dexter's Best Lane house, a bohemian hangout where Yanik had been crashing for the summer.

An older man with a book on raw food and pages of notes sat nearby in the study room, also a listless, scraggly bloke in a green raincoat and rainhat reading the *Guardian*. Seagulls were making a racket outside. The listless man got out a "My Work Plan" benefits form, what sounded like a spray can rattling in his bag (he didn't seem an obvious graffiti writer) and eventually left after donning a large, grubby rucksack. It was getting brighter outside when I left around 3 p.m.

Sitting outside the Boho Cafe peoplewatching, I noted a hipster couple across the street (she carefully delinting him) and was struck by the multiracial nature of the human traffic compared with what it would have been in the late 80s when I'd first arrived here. Max Martin, a keen supporter of the local music scene, stopped to chat and enthusiastically offered me the chance to "take over" an episode of his *Local Hoot* programme on CSR FM. I then saw Juliet again, twice: once passing the Boho clutching awkwardly large bags stuffed with bedding (we acknowledged each other); then sitting near a bus stop in St. Dunstan's Street rolling a cigarette, on her way to Canterbury West for the train home to Ramsgate. On Station Road West she was to run into Glen, a psychonaut friend of Yanik who was soon to abruptly convert to Christianity after an overwhelming experience on the drug AMT.

16:53 to 20:46 on 14th August 2014

Around 6 o'clock I cycled over to Stour Street via Beverley Meadow and the flooded Hackington Place tunnel to rehearse with Tom Holden, a musical project we'd been working on for a few months. He'd recently moved into a house in the Greyfriars estate, rented directly from the Franciscan brothers. The house was still fairly spartan (one room devoted solely to DVDs and computer games) but Tom and his new housemates Toby (his brother) and Tom W were cheerfully welcoming and there was a pleasing sense of a new chapter beginning for them.

I sipped black tea from a Postman Pat mug while Tom smoked out of the kitchen door and we talked about the evolution of computer games, including the soon-to-be-released *No Man's Sky* which he was particularly excited about. This then morphed into a philosophical conversation involving transhumanism, geoengineering and Tom's idea of potential technological expansion so rapid *that it starts to impinge on the past*. Shamans, he suggested, have been accessing *something that hasn't been invented yet*. Terence McKenna's "transdimensional vehicle at the end of time" got a mention. We found that we had strongly divergent opinions about life extension/ immortality and the drive to populate space: he all for it, myself very cautious. Drifting into the front room to join Toby and visitors Oli and James, I suggested that we don't need to venture into physical "outer" space when there's unlimited inner space to explore. I brought up physicist Paul Davies' argument that the Universe is probably some kind of simulation. Tom suggested that simulations could perhaps host the consciousness of things they're simulating. I pointed out how this "computer simulation" idea was very "2014" – in the 14th century there may well have been Franciscan friars on that very spot having similar discussions but using religious language. We've got to that point in history where we're using the computer as a metaphor for reality but why does this seem to fit so well? Tom suggested we're being "invaded from the future" and floated the possibility of bidirectional time (Toby, Oli and James at this point branched off into their own conversation about friends' relationships). I started to expatiate on the connection between biology and the "Arrow of Time", how it's embedded in our language and thinking. Somehow we then got onto the topic of infinity. Toby mentioned having "heard a funny thing saying some infinites are bigger than others", so I quickly summarised the idea of transfinite numbers. We then got onto fractals, the Mandelbrot set in particular.

Tom and I finally started playing music, his tunes "Journey" and "That", and my "Afro-Cornish". James suddenly remembered seeing me playing my saz – a seven-stringed Turkish instrument – in the Greyfriars (now Tom and friends') Garden during the Sondryfolk arts trail in 2011. This led to a discussion about how conservative Canterbury was (but acknowledging the distinction between its institutions and its populus). Tom and I then started playing his composition "Ukelele Thing".

At 8:30 armed police swooped on the city centre after a man was seen brandishing a gun. Two officers with guns were seen near the Old Tannery development. They then jumped into an unmarked car and sped off towards the Westgate. Minutes later they were seen doing a sharp U-turn by the Towers. Police searched the area and arrested 35-year-old Mark Scamp for possessing an air weapon in a public place.

20:46 to 23:51 on 14th August 2014

Tom and I carried on playing while the conversation around us veered through ancient Egypt, the Terracotta Army, a comedic 1968 *Star Trek* episode involving a 1920s-style gangster planet, a Nazis-on-the-moon sci-fi flick and space migration.

Oli lost his iPhone, went into a panic and then found it near to hand a few minutes later. Tom W and Toby headed over to Spring Lane to pick something up while Tom and I played more tunes. Around 10p.m. I cycled over to Club Burrito, the Mexican streetfood place in Butchery Lane that had started hosting live music in its Frida-Kahlo-themed upstairs – I'd seen a poster earlier in the day for a poetry/music open mic there that evening. Carefully weaving around groups of Thursday night drinkers, my besandalled feet were cold from the night air. Quite a crowd outside the Cherry Tree in White Horse Lane. An elegant woman with black pigtails was handing out flyers for The Cuban.

At the bar I ran into another Ollie I knew, a jazz guitarist. He wanted to smoke a cigarette so we sat at an outside table. He mentioned having seen my book *The Mystery of the Prime Numbers* (2010), asking me to summarise it. After initial protests that it couldn't be readily summarised, I gave a very condensed version of the entire trilogy. His girlfriend came out after about twenty-five minutes, furious, having been sitting upstairs alone. I attempted to diffuse the situation by taking the blame for pinning him down with a monologue about mathematics.

Upstairs we caught the last couple of songs from a singer-songwriter called George Oglivie. Bulgarian Yanik (who I'd seen for the first time from the Beaney window that afternoon) then recited a couple of metaphysical poems but I struggled to concentrate due to two drunk, burly tattooed blokes having a shouted conversation in a nearby corner. When he finished, Ollie's girlfriend enthusiastically praised him not for his poetry, but for *his being able to remember it*. The organiser, seeing my saz, pressured me into playing a couple of tunes. I hadn't planned to play, but he persisted, so I decided on an Armenian one called "Ambee Dageetz" and "Deliquescence", one of my own. During the latter, one of the burly drunks got up to leave and surprised me by nodding in my direction with an approving look. Out in the street Yanik was taken by Tom S, a local artist he'd just met, to see a stone he claimed was once part of a "Wiccan altar" (it's actually a kerbstone with a surveyor's benchmark /|\ , originally the ancient druidic "Awen" or "holy wings"). With comic reverence they offered it a splash of Coca-Cola.

I left while blues/jazz/soul duo Claire and Paul were playing, the Argentinian proprietor Luciano insisting on giving me a bottle of Mexican lager for the road (with a slice of lime, naturally), observing politely that my music was "different". Standing outside admiring the Cathedral's illuminated Bell Harry tower, soulful music emanating from the upstairs window, felt like some kind of quintessentially 2014 Canterbury moment. Then, during a brief instant when I'd turned away, the lights suddenly went out on the Cathedral. From down Burgate I heard one side of a loud, drunken mobile phone conversation: "*AWRIGHT... WHERE ARE YOU? WHAT?? WHERE ARE YOU? JOLLY SAILOR... WHERE'S THE FACKIN' JOLLY SAILOR? KEEP WALKIN' UP, KEEP WALKIN' UP... STAYIN' IN THERE OR WHAT?... STAYIN' IN THERE FOR A WHILE... FACK THAT!*"

23:51 on 14th August to 02:19 on 15th August 2014

Opposite Ye Olde Beverlie pub in Hackington I realised that I'd forgotten to put the lights on my bike. I wobbled my way up St. Stephen's Hill, mumbling to myself about my life and friendships, dismounting and getting up on the pavement as cars whizzed by, accidentally crushing a snail underfoot and spontaneously apologising ("*sorry Mr. Snail!*"), then immediately crushing another. A horrible feeling. Noticing that it was two minutes to midnight, I found myself "singing" the Iron Maiden song of that title, despite having no fondness for the band. New steps were being built down from UKC's Tyler Court accommodation block. "Friday 20th June, for 12 weeks", a sign said. Also "CARDY Sorry For Any Inconvenience." The "Any" amused my inebriated mind, imagining Cardy shareholders feeling their share of sorrow for *any* inconvenience *anyone* was feeling *anywhere* in the world. At the top of the hill I ran into the kerb and came off my bike, but remarkably fluidly. I landed on my feet, reminding me of a Taoist proverb about a drunk man falling from a moving carriage unharmed.

I thought back to the night I'd gone to see the celebrated early 90s ska-punk band Citizen Fish at The Maidens Head in Wincheap in 2011 and really did drink too much – cider, of course – getting into the punk spirit, falling off my bike more than once (harmlessly) on the way home. I enumerated what I'd consumed this evening: a discount "World Cup" edition can of Budweiser offered by Oli at Tom's house, half a pint of Heineken and a couple of small bottles of Mexican lager at Club Burrito. "*I really don't feel that drunk, but I do...drunk on Canterbury? Drunk on something very old,*" I mumbled to myself. Some incredible cloud activity was going on, lit up by the rather lovely moon, crickets (or grasshoppers?) oscillating in the hedge across the road.

Back at home I heated up some leftovers, made a mug of camomile tea, set multiple alarms (corresponding to the time divisions in the next few pages), synchronised the clocks in my phone, digital audio recorder and camcorder, and fell asleep hoping for the best.

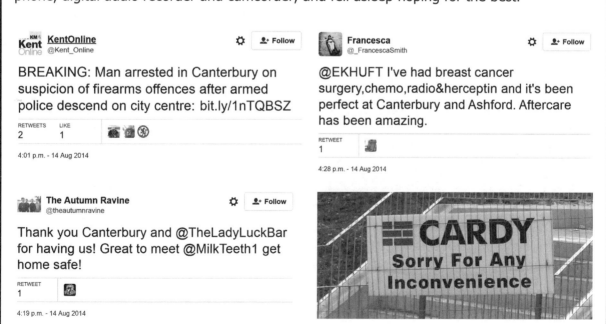

KentOnline
@Kent_Online
☼ 👤+ Follow

BREAKING: Man arrested in Canterbury on suspicion of firearms offences after armed police descend on city centre: bit.ly/1nTQBSZ

RETWEETS 2   LIKE 1

4:01 p.m. - 14 Aug 2014

Francesca
@_FrancescaSmith
☼ 👤+ Follow

@EKHUFT I've had breast cancer surgery,chemo,radio&herceptin and it's been perfect at Canterbury and Ashford. Aftercare has been amazing.

RETWEET 1

4:28 p.m. - 14 Aug 2014

The Autumn Ravine
@theautumnravine
☼ 👤+ Follow

Thank you Canterbury and @TheLadyLuckBar for having us! Great to meet @MilkTeeth1 get home safe!

RETWEET 1

4:19 p.m. - 14 Aug 2014

CARDY
Sorry For Any Inconvenience

02:19 to 04:17 on 15th August 2014

My phone alarm went off at 2:19 as planned. I awoke in deep confusion and flailed around to switch off the alarm but somehow, mysteriously, managed to call Cameron, an anthropologist and patron of the UKC Psychedelics Society. Drifting back into a threshold sleep state, I could hear something I imagined to be a voicemail recording. In actuality he'd awoken somewhat startled and answered his phone, thinking that there might be some emergency involving his mother in El Paso, Texas. It took me 28 seconds of soporific confusion to terminate the call.

My phone (TescoMobile, via the O2 network) would almost certainly have connected with the nearest O2 antenna to where I was sleeping, station 36335 at the top of St. Stephen's Hill, behind UKC's Darwin College. This station would have relayed the signal – the low-resolution digitised sound of a near silent room and my breathing – to Cam's phone (Orange, via the EE network) through at least one other station and/or server. His home being in Rushmead Close, just south of the campus, the signal could well have passed through the nearby Orange network antenna (station KNT0064) attached to the St. Thomas Water Tower on Neal's Place Road.

Mid-Kent Water had bought this tower at auction in 2004 and within a couple of years planning permission for a number of mobile antennas had been obtained. I used to enjoy its architectural elegance back in the '88-'95 era, so on returning in 2007 the ring of antennas just looked *wrong*, like an electromagnetic crown of thorns. It was built by Holland, Hannen and Cubbitts, Ltd. of Holborn, better known for building the Royal Festival Hall and the Cenotaph. Silas Williams, chairman of the Canterbury Gas and Water Company, "opened" it in 1928. More recently, its lower reaches had become popular with graffiti artists as a canvas for their tags and pieces.

Once I'd recovered from the minor mortification of having disturbed someone at two in the morning, I realised that I couldn't remember anything of my dreams. I went back to sleep, having vague dream impressions of being back at Club Burrito, some unorthodox arrangement of tables in the street... I later deduced that for that call to have been made, it would have taken at least eleven keypresses on my part or else something highly anomalous to have occurred within the phone's circuitry. Was it hypnagogic dialling? Pixies?

04:17 to 05:51 on 15th August 2014

The next alarm went off. I couldn't remember much but it felt like my disrupted dreams were being shaped by the idea of fitting everything around a timeframe.

Asleep again.

I dreamed that I was sitting in a cafe in town. Various people, none familiar, were petitioning me to include various things in this book. These all seemed a bit tangential – city councillors of the past and their various artistic endeavours, something about a snowball fight – stuff I felt didn't fit. I was trying to get people to spell names, getting confused about who was who.

5:51 to 7:06 on 15th August 2014

I was woken by yet another alarm and did my best to remember some very vague dream content. Birds were singing. Still very sleepy, I decided to snooze a bit more. I was feeling good about things and glad to see that it wasn't raining.

By 6:15 I'd not really managed to get back to sleep, was just floating on the threshold. A couple of insect bites were making me scratch. It was misty and overcast but felt like it could brighten up. I could hear an altercation involving a couple of noisy birds.

I drifted off again and dreamed of a working class family living in a house in Northgate (seemingly in the 1950s) being sold the idea of the slum clearances – mum, dad and son. A rolled-down banner was telling them in black capital letters that colour TV and hot running water would change everything. They were considering it.

I woke again just before 7 o'clock and put on Radio 4 to hear the headlines. I somehow missed the weather forecast. Nouri Al-Maliki had just resigned as Prime Minister of Iraq.

7:06 to 8:06 on 15th August 2014

Although basically awake I was still cruising in and out of various planes of consciousness, listening to wind in the trees, the rumble of aircraft, wood pigeon coos and chirpy bird sounds.

I decided to make tea and read in bed for a while. At 7:30 I was reading Richard Feynman's account of the "double-slit experiment" in quantum mechanics for the third time but struggling to focus due to being shattered by broken sleep. I was also feeling some mild anxiety about how this morning would go. I was trying to engage with the concepts of constructive and destructive interference in wave dynamics but my mind kept going off on tangents. The sky was still grey, wood pigeons now dominating the soundscape. I resolved to get up, stretch and make porridge.

Just before 8 o'clock I put the radio back on and this time caught the weather. It was supposedly going to clear but with possible heavy showers. An "autumnal kind of day". I decided to wear long trousers and sandals for my walkabout but to carry shorts in case it became uncomfortably hot. I was aiming to look as nondescript as possible. Picking blackberries not far from my front door I spotted a cluster of St. John's wort flowers. No one was about.

TIMES **Times Series**
@TimesSeries
⚙ Following

The Police are warning people to keep their mobiles safe after a spate of thefts in Canterbury canterburytimes.co.uk/tips-protect-m ...

11:35 p.m. - 14 Aug 2014

8:06 to 8:53 on 15th August 2014

I cooked porridge as I assembled the various things I needed for the morning. Thinking about my camcorder's battery life and how I was going to organise the day, I realised how self-referential my account of this morning was already becoming. I also noticed how my minor anxieties were affecting my coordination and so made a conscious commitment to be careful cycling and walking, not to bump into people. Spatial awareness wasn't my strong point at the best of times! I was a bit all over the place and I knew it. Something about the Bolsheviks' exposure of the secret 1916 Sykes–Picot Agreement was being discussed on the radio. I headed down into town at 8:30.

Someone had dumped a load of large paper napkins at the top of St. Stephen's Hill, some kind of cranesbill or mallow growing nearby. A black car with a multicoloured replacement door passed. Various people were coming up the hill, walking, running and cycling: a woman in a pale green T-shirt jogging; two awkward, overdressed young men; two black guys; a young oriental man listening to an iPod; an oriental couple, she eating something out of a paper bag. An Asian man in a red shirt heading down the hill on my side took his jacket off and started running, really going for it (not a jogger). A group of youngish people coming out of Hales Place struck me as probably foreign UKC postgrads. I started thinking about the historic traffic up that hill relative to what was now going on demographically. A prominent "OREO" tag remained on the postbox at the corner of Manwood Avenue. I passed the red-shirted man who'd now stopped running. A "Keep Beds in Canterbury Hospice" sign was on display in a garden. People were generally dressed like it was October – understandably, considering the weather.

A 21A bus headed towards the city centre, then a number 5 to Chestfield and Seasalter went by the other way. A highway maintenance worker in a pickup truck parked near the entrance to Beverley Meadow appeared to be reading. Two magpies, a few seagulls and a couple of pigeons milled about near the footpath. A man on a mobile passed talking Italian. Crows (or rooks?). A woman eating citrus fruit. A black man with a strong South London accent on a mobile headset: "...but then, the way I see it..." Little strips of blue were appearing in the sky as I approached the Hackington Place tunnel. The bag on my back felt awkward, needed repacking. A discarded beer bottle in the tunnel. A tiny bird (a wren?) fluttered across the path. I turned right down Station Road West and saw a little silver sports car with the number plate "OOH BOX" and a person looking downtrodden — or perhaps just looking at the pavement? The Goods Shed doors were open, fruit and veg all out on display but no humans visible. A sign announced "Fresh Fish at the Goods Shed/open 10-7 daily/from Broadstairs/Great Prices".

I locked my bike behind Canterbury West station. A woman sat alone on Platform 1. On the plaza out front an androgynous young woman speaking Dutch on her mobile was wearing a T-shirt announcing "I don't need sex, the government f**ks me every day". Not much was going on inside the ticket office – I'd missed the work rush. Those flowers for the child killed in Gaza were still tied to the railings. A parking ticket machine was covered in graffiti tags. I was now starting to feel self-conscious about my secret filming plan.

8:53 to 9:31 on 15th August 2014

Outside the betting shop at the St. Dunstan's end of Station Road West, I saw Angus, a kind of unacknowledged street-level Archbishop of Canterbury and outsider installation artist. He was rolling a cigarette on his guitar case while addressing two associates (one who'd half-heartedly blagged me for change outside The Beaney the night before): "...*in that order...*" He had their attention, as if giving instructions. Two young Turks stood outside the Turkish barbershop across the road. Meandering through the Westgate Gardens I stopped to look at a stone memorialising Daniel "Taihg" Lloyd. Then another memorial: "Michael John Burt: His Friends Miss Him". I suddenly felt for everyone left behind by anyone they've loved – it struck me that living almost always leaves behind a trail of sorrow. Posters in the noticeboard announced free community events: an archaeology dig later in August and wild food foraging in September. Budleia was growing from the wall of the Ocakbasi takeaway, once a restaurant called Fungus Mungus where I would sometimes wash up in the evening for a free meal and a lift back to Whitstable with Nick Dent in early '93. Down St. Peter's Lane, clocking "Too Weird" and "DQ" graff tags. Workmen were digging up the road on the corner of Westgate Hall Avenue as a young woman emerged from a house tapping daintily on her iPhone. I found the back entrance to "Solly's Orchard" and read a plaque explaining that it was once a 17th century orchard on the site of a former Blackfriars gatehouse.

I'd decided to head for the Cathedral, opting for the familiar route via Mill Lane. Sunshine! In Blackfriars Street I stooped to connect with a grey and white cat, very intelligent-looking but completely uninterested in me. The stonemason from the day before was again at work on the 13th century refectory. A young man called up to him in an Estuary accent, *"Morning! You got the keys to the room upstairs?"* Starting to feel a bit doubtful about this whole project, I caught a glimpse of a daring builder on a King Street rooftop bounding over eaves with a load on his shoulder, someone (him?) singing in a falsetto soul-type voice. Roadworkers outside St. Alphege's Church clutching Costa takeout coffees were mumbling and swearing at each other. A peculiar, sustained sound was being made by a man encircling a large cubic stack of Belgian beer crates with black vinyl tape outside La Trappiste.

Canterbury Cathedral
@No1Cathedral

⚙ Following

On the Feast Day of the Blessed Virgin Mary, let us pray for peace throughout the world and the coming of the Kingdom

RETWEETS 2   LIKES 2

1:29 a.m. - 15 Aug 2014

Tourists were snapping photos of Christchurch Gate as usual as a big blue Viridor rubbish lorry reversed into the Buttermarket. I took a quick look at the recent WWI memorial display then showed my local Residents' Card at the gate and wandered in. It dawned on me that the Cathedral precincts took up a considerable proportion of the old (walled) city so I resolved to spend a correspondingly significant part of the morning checking it out properly. Seagulls squawked overhead as I approached the Cloisters and contentedly noted an old postal wallbox set into some worn brickwork. Wandering around the Cloisters, I was deeply surprised to hear the sound of clucking hens coming from behind a chunky oak door next to the library entrance. What?

9:31 to 10:01 on 15th August 2014

Moments later an affable-looking clergyman walked past so I just had to ask: "*Is, is there supposed to be... a bird inside... there, do you know...?*" "*Yeah, yeah, yeah, it's a h..., they're hens,*" he replied. OK, hens. But why behind that door? I later found out that behind the wall lies a small kitchen garden and that the Dean keeps hens back there. But at the time all I could think about was Voodoo sacrifices.

Nearby, a very thin gardener with dyed red hair and a wheelbarrow put her gloves on as I decided to walk a loop around The King's School's Green Court. Latin dedications to the dead stood at the bases of trees: "*FREDERICVS CLEARY CAVSA HVIVS SOLITA VIRIDATATE CONSERVANDAE*" (MCMLXXVII). At the far corner I found an inviting stone plaza-like space with a war memorial and a Norman staircase. Its powerful atmosphere left me feeling like I shouldn't be there although there were no signs to that effect. There was something beautifully proportioned about the stonework. Some workers sat against a nearby wall eating takeaways.

Onwards. I attempted to study a wall-mounted map which reconstructed the medieval priory but couldn't process it — too much information. "*It is requested that Persons will not deface the walls by cutting or writing thereon. By order of the Dean and Chapter.*" This hadn't prevented a palimpsest of historical tagging: "Bryner Was 'Ere 1980", "Peter", "Nehdi"... I was feeling a real sense of peace in the place but reminded myself that there'd been a lot of different things going on here for 1500 years — rather like Jerusalem, some of the best and some of the worst of everything mixed together. I found the herb garden: sage, wild marjoram, apothecary's rose, hop, dog rose, blackthorn, hawthorn...

Finding a little statue of St. Dunstan holding the Devil by the nose with a pair of tongs, I was amused by the way Dunstan is casually raising his other hand looking as if to say "No worries, everything's cool." I was hit by a wave of memories of the Cloisters: meeting Laptop the Cathedral cat with my visiting friend Stella on my birthday in 2012; lovely Vicky singing furtively in the Chapterhouse in 2004, almost appearing to float slightly above the floor; Dave "Nusphere" Prentice and I recording some overtone chanting and saz there one evening a couple of years later.

Passing through a door into the Crypt I headed straight to the Chapel Of Our Lady Undercroft but there was a grumpy looking man sitting on one of the chairs there so I decided to wait for him to leave. I wandered around for a while, taking in a statue of St. Thomas gifted by the Church of Sweden, an Anthony Gormley sculpture and some large columns brought from the Saxon church of St. Mary's, Reculver. The man was still there but I wasn't *that* bothered so I put 20p in the box and lit a candle before sitting and meditating on Mother Concordia's beautiful sculpture of Mary and Jesus (1981). After a while I started to experience some unsettlingly weird optical phenomena. I'm sure the effect of the gently flickering candles on the shadows was largely behind this but, even so, I've always felt a sense of power in this place. A slight fear of slipping into madness kept pulling me out of my trance despite my attempts to trust and just go with it. An alarm that I'd forgotten I'd set for 10:01 went off, leaving me slightly embarrassed and flustered.

10:01 to 10:25 on 15th August 2014

Someone else's (fairly placid) ringtone went off nearby, both annoying and reassuring me. At 10:06, mine went off *again* – I must have accidentally hit the snooze button – even more flustering than the first time. I dropped back into a trance state for a few more minutes and left the Crypt at 10:13. On my way out I stopped to read about some stained glass illustrating Mary's descent from Nathan, son of David, a detail of divine royal lineage I was unfamiliar with.

I headed towards the Trinity Chapel noting the very different energy up there and resolving to attend Evensong sometime soon. A flock of tourists with earpieces passed through the Quire. I stopped in to look at the tomb of St. Anselm and found myself wondering what it would've been like to have lived in a nearby village during the peak of the pilgrimages – how walking to Canterbury wouldn't have be anything out of the ordinary.

I peered into a vestry and saw a couple of people getting robed up, then moved on to look at photos of WWII bomb damage to the Cathedral Library and at the painting depicting the legend of St. Eustace. The latter formed the basis for the "Eusa" cult in Russell Hoban's extraordinary 1980 post-apocalyptic novel *Riddley Walker* (set in the area around "Cambry"). A group of four women in plain, unfashionable skirts and headscarves walked past. American, I guessed. Nuns? Mennonites? I didn't recognise the style. I remembered wandering around the Cathedral while interviewing Gong frontman Daevid Allen in 2012 and decided to go to the gift shop to buy a Second Commandment fridge magnet which we'd laughed about ("*Thou shalt not make unto thee any graven image.*") Helpfully identified by his nametag, I was served by Matthew P.

"*Would you like a little bag for that?*"
"*Yeah, why not?*"
"*Nice little Cathedral paper bag. £2.50, please sir.*"
"*Could you tell me the significance of the 'IX'? I've been wondering about that.*"
"*It's the emblem of the Dean and Chapter. The letters stand for 'Jesus Christ' in Greek.*"
"*Ah yes, Iesus Xristos?*"

Having noticed a few examples of this insignia around the place I pointed out that in some instances the "I" is written in upper case, in others it's lower case: "IX" versus "iX". Matthew P explained that the Cathedral authorities couldn't make up their minds – they were still looking for a definitive early example. He preferred the upper case version. I suggested that the lower case "i" seemed a bit more quirky, even modern (as in "iPhone" or "iPad").

10:25 to 10:45 on August 15th 2014

Realising that I'd overlooked the Black Prince's tomb, I wandered back up to the Trinity Chapel for a look, triggering memories of cleaning up the "Black Prince's Well" at the 11th century lepers' hospital in Harbledown with my friend Tim some years earlier. I paused to look at the tombs of Simon Sudbury (the Archbishop killed by revolting peasants in the 14th century) and Henry IV. Henry's body, according to rumours, may have been dumped in the Swale by superstitious boatmen or hidden in the church at Chilham.

Leaving the Cathedral, I read that the men of St. Leonard's, Hythe would be singing Evensong at 5:30. I decided to exit the Precincts through the Queningate, something I hadn't done before. Looking at some beautifully reconstructed stonework high up I thought back to a conversation at a party in Boughton with Rob who'd briefly apprenticed as a stonemason here and had much to say about the Cathedral, its history and influence. To get through the Queningate involved walking through the Kent County War Memorial Garden but there was some sort of service going on there – men in military uniform, women in big dressy hats, wreaths, someone in a busby-like hat with a silver trumpet. I was curious to have a look but the entrance was guarded by two Cathedral "bouncers", something I'd not before encountered.

me [politely]: "*What's the service going on here?*"
bouncer [reluctantly]: "*For the Men of Kent. It's the annual service they have every year.*"

So I instead left the Precincts via an entrance out onto Burgate that I'd never noticed before, then got out my calculator to start generating random numbers to direct me around the city. A slightly better-off class of alcoholics than the street-drinking types were sitting outside The Thomas Ingoldsby with their mid-morning pints. This pub had recently become part of the Wetherspoons chain, its name memorialising a character created by local comic author and practical joker Richard Harris Barham, now largely forgotten. Barham was baptised in nearby St. Mary Magdalene church, only its tower now standing across the street. The first batch of random digits sent me up along Burgate Lane towards the site of St. George's Gate. Outside the "Strict and Particular" Zoar Baptist Chapel (1845), a pair of middle-aged British tourists in flat caps read aloud from a guidebook that it used to be a water tower. That had been after the 1801-2 demolition of St. George's Gate whose towers had housed until then the city's water supply.

Not much going on outside KFC. I checked the memorial slab in the floor of St. George's Clocktower – not a memorial to Christopher Marlowe as I had misremembered, although he was baptised in that church. Up the taxi rank on Canterbury Lane. I found a "Vomit Here" stencil on a dark green utilities box on Link Lane. Down Iron Bar Lane. The numbers were taking me back down Burgate. I stopped to look at the tower of St. Mary Magdalene – the church itself was demolished in 1871, a curious memorial obelisk remaining behind glass. Nearby, a shop was selling tacky Thai trinkets along with some beautiful clothes. I was struck by a couple of gargoyles I'd failed to notice before on the Tudor building at 67 Burgate – a lovely little sweetshop/newsagent. I stopped in to scan the newspaper headlines – all but one involved the latest Cliff Richard allegations.

10:45 to 10:59:58 on 15th August 2014

Sticking my head into the Polish food shop on Burgate Lane I wanted to ask its proprietor about the Polish community in town... but not wanting to be misunderstood in the current anti-immigration climate I opted not to. The random numbers led me back down Link Lane. I felt mildly annoyed that the path being generated was ugly, nondescript and not particularly "historical", but then decided that the backs of shops I was seeing represented the "guts" of commerce rather than the attractive face presented to the shopping public and so were more "real" and interesting in some ways. Back onto Burgate. I recalled reading about a Carver & Staniforth's bookshop somewhere along here which was destroyed by a bomb in 1940 along with Miss Carver and her cat. She'd been known locally for serving tea to favoured customers in a back parlour.

Past the Cathedral giftshop and into the Buttermarket, I stopped to peer into the Cathedral Gate Cafe, thinking of a story I'd heard about a little old lady protesting the closure of the tea rooms there in the 80s, the unexpected marijuana reference made by an American GI having tea there in the 1944 film *A Canterbury Tale* and its recent, almost heretical, tenure as a Starbucks. Tacky souvenir shops were all around, easy money for the Dean and Chapter who owned all the buildings here. They'd suffered heavy losses in the 2008 Icelandic bank collapse, hence the profusion of these. Down Mercery Lane, I quickly inspected some newish carved gargoyle buttresses and was then steered right down the section of the high street called "Parade". Seagull sounds. Into Best Lane Garden. No drinkers, unusually. Happy memories returned of the "arts trail" feast there in 2011 with beautiful acoustic music from Liam and Raven in the evening sun, also recollections of an account of five musicians being paid a shilling each to play loudly from the top of All Saints Church (this having once been its churchyard) for the visit of James I in 1616. Since 2010 it's formally been the "Three Cities Garden", celebrating a civic link with Vladimir in Russia and Bloomington-Normal in Illinois, USA which began in '85, a Thatcher-era cold war legacy.

Passing the Thomas Becket pub, its ceiling strung with hops, I recalled a curious evening there in 2011 chatting to a visiting preacher from Denton, Texas who was making origami insects and liked my description of the city as "a palimpsest". Immediately afterwards I'd cycled straight into a kerb beside the Marlowe Theatre and ended up crumpled and moaning on the pavement, seemingly because I'd been seeing the road layout not as it was, but *as it had been* before the new Marlowe was built. Back in the present, a young boatman punting tourists up the Stour was getting carried away with his storytelling while a couple of young Asian men looking at a smartphone in the riverside Friends' Garden talked loudly. I perused the Marlowe's poster display of various populist events in the near future, then considered the recently installed life-sized sculpture of pantomime actor Dave Lee sitting on a bench. A sign outside the Dominican Friary Community Centre (the other surviving Blackfriars building, also 13th century) described it as "Formally [*sic*] the guest house for pilgrims 1315-1538". Bells chimed from the Cathedral.

Kent Cricket News
@ksncricket
⚙ 👤+ Follow

Hampshire have won the toss and will bat first here at Canterbury against Kent in the County Championship game. #superkent

2:46 a.m. - 15 Aug 2014

10:59:58 to 11:12:09 on 15th August 2014

Down St. Peter's Lane, I realised that I'd got quite a pace up, driven by an urge to just keep moving. Part of this had been a fear that people might notice that I was filming with my hidden camera-in-a-shoebox, but I'd developed a technique where I could just turn so that the side of the box faced them, defusing any such paranoia. I was starting to enjoy this experience now, felt grateful that it wasn't raining and decided that the random numbers were a good idea after all. Down Pound Lane I noticed a semicircular tower at no. 16, once part of the city wall but now an architects' office — somehow I'd never really noticed it before. The former presence of the wall along this route was suddenly made more real to me by this discovery.

I was led into the "butterfly garden" by the Stour, which a plaque told me was planted in memory of local teacher Kenneth Pinnock, author of *A Canterbury Childhood* (2010). Wood pigeons were audible. I'd always been confused by what the river did around here — multiple branchings. Walking along one of them I found myself admiring the little back gardens of the houses of Blackfriars Street, wondering what it would be like to live in one of these. Litter caught up in river weeds. Someone had left their jumper on the ground. A "New World Order" sticker on a lamppost featuring eye, pyramid and "2013", also a "WAKE UP NOW" sticker. Over Guy's Bridge (installed in '94), I walked along Pound Lane again and into the car park, wondering why the Cathedral bells were now ringing so much. Women's laughter from somewhere. I felt somehow "wrong" being in a car park, but then imagined what used to be there (probably dozens of little houses). Although flat and paved now, I contemplated how it wouldn't take long for vegetation to start pushing up through cracks and return it to wilderness. Temporary.

The "secret" garden I'd found a few days earlier seemed pleasantly homemade — scaffold poles, palettes, etc. It almost felt *unauthorised* but it was hard to imagine how that could be. I was curious to know who was behind it. Someone had put a lot of love and work into making this place beautiful. The fruit trees triggered images of orchards, Kent as the "Garden of England" and how beautiful a whole *county* could be if it were this well loved.

 **Canterbury Cathedral** @No1Cathedral   ⚙ Following

Men of Kent & Kentish Men, Band of Brothers & Old Stagers remembering the dead of 2 World Wars at @No1Cathedral today

3:12 a.m. - 15 Aug 2014

Wandering on, I considered stopping somewhere for a cup of tea but decided that random walking was the trip I was on so I should stick with that. Children's voices. I turned right, feeling a bit hungry now, thinking about how the mundanity of the city was no longer an issue — *everything* about it was interesting in the self-induced altered state of consciousness this project had led me into. Up onto the Marlowe Theatre steps, ladies could be seen taking tea behind glass.

11:12:09 to 11:21:52 on 15th August 2014

More seagulls. I drifted past the Christopher Marlowe memorial featuring the topless muse holding an unstrung lyre. She was once known to locals as "Kitty Marlowe" and had been moved from the Buttermarket (being considered too risqué in Victorian times to face the Cathedral Gate) to King Street (to make way for the WWI memorial) to the Dane John Gardens (where she was damaged by a WWII bomb blast and vandalised in 1977) and then to here in 1993 when she was rededicated by the actor Sir Ian McKellan. A pair of young employees of the "Canterbury Tales" multi-sensory tourist attraction in St. Margaret's Street walked past dressed in awful pseudo-historical felt jester gear. Moving on down The Friars, I noticed for the first time a small locked gate into St. Peter's churchyard, installed in 1990 to memorialise the "Huguenot forebears" of one Wendy Legrand Allday, "many of whom are buried here". A burble of voices, a clatter of scaffold poles, fragments of overheard conversation: "*...hundred-year-old man has...*"

Down St. Peter's Street past The Lady Luck. In the early 1960s when this pub had been called the Three Compasses, future Soft Machine legends Kevin Ayers, Robert Wyatt and Hugh Hopper were once ejected for having long hair. These days it's a "subculture" pub, welcoming hippies, punks, goths, LGBTQ types and just about everyone else. A lorry engine roared to life beside me. A poster announced something going on at the pub: an all-day memorial gig called "Luckfest" for a young barman called Luke Carry who'd died of leukaemia in March. Stopping by a low wall in Black Griffin Lane to scribble down a sequence of random digits (to save from having to fiddle with my calculator at each junction) I noticed a "roached" packet of Rizlas someone had forgotten and an empty beer can with the four-pack plastic binder still attached. Lots of 9's. Deciding to keep moving and suddenly striding onward, I slightly startled a little old lady in a pale blue coat. The Methodist church brought to mind something I'd read about John Wesley's visit in 1764 to open the original "pepper pot" chapel in King Street. A lot of graffiti tagging around here.

Back on St. Peter's Street I wandered past the Army recruitment office, thinking about how they recently got both an IRA letter bomb and "bombed", in the graffiti sense, with purple spray paint. The retro clothing shop "Funky Monks" brought to mind the profoundly un-funky radical breakaway 13th century sect known as the "Friars of the Sack" whose HQ was a few doors down. I wondered when Dave Radford's seemingly eternal record stall had left the Indoor Market. He'd been in a local mid-70s prog band called Gizmo and Hugh Hopper sometimes used to mind his stall. Next to the city's coat-of-arms above the door of the Sidney Cooper Gallery I spotted a circular "Mary Tourtel studied here" plaque. Back down Pound Lane: Barretts car showroom (since 1902, now suppliers of the Lord Mayor's official Jaguar, also recently graffiti bombed); a sculptural sun hanging in front of Cafe du Soleil.

University of Kent
@UniKent

⚙ ➕ Follow

Got a place through Clearing? Come to our open day tomorrow to see your new home #hellokent

RETWEETS LIKES
5   2

3:29 a.m. - 15 Aug 2014

11:21:52 to 11:29:37 on 15th August 2014

I walked along the Stour past a spot where there were usually a few people drinking Special Brew but none were to be seen. Three left turns. I realised that I was going to loop back on myself. I wasn't keen on that but it was what the numbers were dictating. The sound of something beeping repeatedly mingled with the general hum of traffic noise. Back down Pound Lane *again*. "Katie 4 Jacob B" written in chalk. I was getting thirsty but resolved only to get water if I passed a place where it was immediately available. Back over Guy's Bridge, then right towards the butterfly garden again. Back past the discarded jumper, following the Stour. Seeing ducks on the river triggered the Incredible String Band's metaphysical song "Ducks on a Pond" (1968) in my mind, bringing to mind the fact that former Archbishop Rowan Williams had written a foreword to *beGLAD*, a 2003 compendium of their songs (this had convinced me that he must be basically OK). A flowerpot floating in a broken wicker basket.

The green space beside The Causeway had for some years been home to imaginative sculptures made from the root balls of trees brought down there in the Great Storm of October 1987. Still quite new when I was a student, they were left to decay gracefully and eventually removed. I noticed that The Marlowe Theatre had some kind of studio space in Pound Lane, formerly the (tiny) temporary library space when the Beaney Institute was being revamped. They were being a bit more proactive in promoting local theatre-related talent – good to see.

I stopped to look at the caged Judas tree planted across the road from the Millers Arms in 2012 to commemorate the thousandth anniversary of the murder of Archbishop Alphege the year after the devastating Viking siege of Canterbury. Across the road I took a closer look at the *vesica piscis* St. Radigund painting above the door of no. 8 St. Radigund's Street.

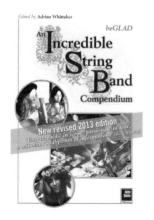

Christ Church SU
@christchurchsu

⚙   👤 Follow

Can't wait to see the new @csrfm website! It's being launched at midday today csrfm.com!

RETWEETS   LIKES
3   2

3:25 a.m. - 15 Aug 2014

Live Cricket Score
@criclivescore

⚙   👤 Follow

Kent vs Hampshire 6-0 #cricket

3:24 a.m. - 15 Aug 2014

11:29:37 to 11:35:47 on 15th August 2014

Past The Parrot, I stopped to look at the nearby chunk of original Roman wall (apparently one of the best preserved in England), thinking of the evening in April when I ran into the members of Arlet eating chips beside it before a gig. The pub had been called "Simple Simons" during my student years. My first visit was for an inaugural meeting of the UKC Graveyard Appreciation Society but, finding them a bit annoying and gothy, I'd decided not to join. Around the same time I'd considered joining the Church Bell Ringing Society, but for some reason talked myself out of that. As an undergraduate here I'd only got involved with UKC Radio (recently absorbed into CSR FM) and the university orchestra (playing French horn), never joined a proper society.

Down the very narrow Church Lane, which I'd only recently discovered, I came out on The Borough (its name reflecting its course along the perimeter of the original Saxon *burgh*) alongside St. Mary's Northgate. A woman's voice: "*Well I didn't want to sort of...*" I crossed over towards The Jolly Sailor, taking me back to the night when Syd Arthur played in a marquee in the pub garden, being mistaken for a homeless beggar as I sat in the street listening, wrapped in a blanket. I turned to look at 5 St. Radigund's Street where Steve Hillage and Spirogyra had lived over 40 years ago, also my Sondryfolk friend Laurie's mum Angela in the late 70s when what was now The Parrot had been a printing workshop for the art college where many of her friends studied. Just off New Dover Road, this later became Kent Institute of Art and Design – Laurie studied there – and most recently University of the Creative Arts. On the corner, Skunkworks was still trading, the headshop now under the scrutiny of MP Julian Brazier as a result of Taihg and Hugo's tragic MXE-linked deaths and another youth almost dying on a "research chemical" sold there. The random numbers were directing me to walk one of the least pleasant stretches of the city wall route: down Broad Street. A small Elizabeth II souvenir automaton was grinning and waving uncannily in someone's front window. I couldn't help thinking about Jimmy Savile and how he'd managed to get so close to her husband and eldest son. A bloke with a rucksack gobbed while crossing Albion Place. Noisy traffic. I stopped to check my camera battery near The Victorian Fireplace.

 **Police in Canterbury** @kentpolicecbury ⚙ 👤 Follow

Hi everyone, PCSO Hinwood and PCSO Bolton on #mobilepolicestation today. First stop #canterbury high street at 12. See you there!

LIKE 1

3:33 a.m. - 15 Aug 2014

Kent Union @KentUnion ⚙ 👤 Follow

@KentEssentials have household and health & beauty products on sale in store, to make room for the freshers' stock. Grab a bargain!

RETWEET 1

3:35 a.m. - 15 Aug 2014

11:35:47 to 11:40:42 on 15th August 2014

I paused to inspect some old bits of fireplace left outside the shop. A kid whizzed past on a BMX bike, cycling on the pavement (tsssk!). Just along from there, the Canterbury Archaeological Trust office had chunks of historically significant stone in its front garden and a memorial to CAT gardener Margaret Cowles (1919-2007) attached to a wooden fence.

The traffic was making this route pretty unpleasant but I reminded myself that I'd soon be elsewhere. And that I was following the path of the old city wall. The wall itself came into view beyond the Queningate Car Park. Waiting for the lights to change at a pedestrian crossing, a bearded man in a red top with a "granny trolley" and two little boys snapped at one of them *"Stop it! You're going to hurt somebody with it!"* Looking across to Lady Wootton's Green, I could see the Abbey's Fyndon Gate but the statues of Ethelbert and Bertha weren't quite visible. I thought of Bertha and how the Queningate is believed to have been the "Queen's Gate" through which she'd have walked each day from the royal palace to worship at her private chapel, now St. Martin's Church. On the inside edge of the car park I stopped to look at a historical information board but again found it too much to take in. A nearby voice asked of someone *"Where's your credit card?"*

Next to the red phonebox at the end of Burgate a recent piece of stencil art read *"Capitalism: The Road to Hell is Paved with Profit and Loss"*. Down Burgate Lane past the Polish shop, I glanced in to see the shopkeeper inside eating a roll, presumably with something Polish in it. On past the "Money Shop" offering dubious loans and Western Union money transfers.

11:40:42 to 11:44:37 on 15th August 2014

I passed Zoar Chapel again, properly noticing for the first time the new "contemporary Asian" restaurant Chom Chom across from it. A new "LONG ER" graffiti tag had appeared on a door next to the cash machine and phonebox at the end of St. George's Street. Crossing over towards the bus station, a burble of voices: "...*very long...*", "...*I'll open it...*"

A lot of bus noise, a man idly looking at a toy display in a window, a woman talking to a child in his pushchair: "*I knoooow*". The numbers were sending me into the Whitefriars shopping centre. A display board with a map of the city centre had "visit Canterbury" written in the corner with "Canterbury" in the intentionally messy font that the city authorities had recently adopted as some kind of branding exercise. Thinking about fonts, I remembered that there's one called "Canterbury Sans", something I'd discovered when Googling the name of my monthly music podcast *Canterbury Sans Frontières*:

A voice: "...we go over there." I was struggling to remember what I'd read about the namesake Augustinian friars who were based around here. They were the closest of the local religious orders to the medieval trade guilds, the commercial sector of city life at that time. But they probably would have been shocked by the La Senza window display featuring provocative images of women in lingerie. Two little girls ran past, their footsteps echoing off the high brick walls. Images from archaeological maps of the medieval Whitefriars area were engraved in the paving stones, a typical "heritage" concession to what was once here. A stall selling incense and Buddhas. People sitting on a wooden gazebo-like structure, some texting, one looking at me slightly suspiciously. Voices: "...weather's not great but...", "No, the weather's no good..."

11:44:37 to 11:47:45 on 15th August 2014

Down Gravel Walk and left onto St. George's Lane along the edge of the bus station. A voice with a strong Estuary/Kent accent: "*So, I'm having to grow up...*" A man passed wearing earphones, a garish yellow plastic bottle swinging in his right hand. A mother and little daughter with shopping bags presumably running for a bus rushed across the zebra crossing just ahead of me. The beeping of a reversing vehicle, the rumble of buses. People sitting under the arcades waiting to board their Park & Ride bus. They'd removed the irritatingly imperative slogan "Love Your City" from these, I was pleased to note. An older lady with a turquoise handbag checked her watch. I stopped to look at the bronze statue of a lamb standing on a tree trunk – it was by Kenny Hunter, just called *The Lamb* (Whitefriars' online blurb explaining: "*The bronze work addresses a diverse range of historical, religious and contemporary social currents... This organic, pastoral composition contrasts strongly with its location... inspiration for the work includes William Blake's 'Jerusalem', the hymn being synonymous with English identity and yearning for social justice.*")

Past a kid wearing an Oakland Raiders top and tracksuit bottoms advertising Duff (the fictional brand of beer from *The Simpsons*), then a mum holding one child's hand, another under her arm, beside a man pushing a pushchair. Round the corner, back towards the Clocktower (all that was left of St. George's Church after the Blitz), a kid passed in a "Criminal Damage" T-shirt. The numbers were going to take me left back into Whitefriars, disappointingly. As I reminded myself out loud to "be present", one of a group of tourists on the corner of Rose Lane (for some reason I guessed they were Korean) said something involving "KFC".

11:47:45 to 11:50:13 on 15th August 2014

Ahead of me, someone in a kiosk was selling Sky TV packages. "2 YEARS FREE" promised the sign. "*OLLIE!*" shouted a child amidst the babble of shoppers' voices. I thankfully got to turn left down an unnamed lane back out of Whitefriars towards the bus station again. "*Would you cry?*" a young man asked another. Right on St. George's Lane past a pigeon pecking crumbs under the arcades. Approaching Boots the Chemist, a little girl in a pink coat: "*...not going to be raining...*"

I passed a silver taxi with the phone number "666 666" prominently displayed, something that always amused me in this ancient Christian city. Easy for drunk students to remember, I supposed. Moments later, a silver van from the Whitstable-based plumbing company "MAYDAY!" passed in front of the gates of the Dane John Gardens. This took me back to the Whitstable Mayday celebrations in 2009, watching a proliferation of diverse Morris dancing "sides" with my friends Andy and Tim, we cheerfully joining in the singing of a neo-pagan Mayday song in the presence of a "Jack-in-the-Green" (a balding middle-aged Morris dancer hidden inside a large conical mass of foliage – we stuck around to watch him climb out, an oddly disturbing moment). Turning right on Watling Street, the ancient Roman road and original main thoroughfare through town, the distinctive cupola of St. Andrew's United Reformed Church came into view.

 **The Sewing Shop**
@SewingShopSunSt

Sale now on! p to 70% off fabric, 20% off Moda patchwork packs, and 10% off other fabrics/yarns/trimmings...

3:48 a.m. - 15 Aug 2014

11:50:13 to 11:52:13 on 15th August 2014

I remembered that Watling Street had entered the city through the old Ridingate, now just a nondescript 1970 footbridge joining two sections of the city wall. Right down Rose Lane. The randomness was leading me back into bloody Whitefriars again! I despaired, momentarily stopping to parse the Whitefriars logo inscribed above the entrance to a parking garage. After a moment I realised that it was a stylised "W" created from three adjacent arches (two only half-visible), subtly invoking medieval religious architecture.

A reversing vehicle pierced the air with its repetitive beep as a couple of people whizzed past on mobility scooters. I realised that I was on Rose Lane, remembering how a Rose Lane Garage had featured in the 1944 film *A Canterbury Tale*, the character Alison Smith (*below centre*) having her caravan stored there and learning from its proprietor, the wonderfully named Mr. Portal (*below right*), that her missing serviceman boyfriend had turned up alive in Gibraltar. I'd recently read about the 1951 "Canterbury Exhibition" held here as part of the Festival of Britain, vaguely recollecting something about workmen wantonly demolishing the ruins of the Whitefriars friary when removing the exhibits.

11:52:13 to 11:53:48 on 15th August 2014

I was approaching Waterstones bookshop, one of the few places in Whitefriars where I'd ever bought anything. Past Carphone Warehouse and The Body Shop. Voices: "...*you need any help*...", "...*work tonight*...", "...*all 'round Canterbury*..." Two older women walked next to each other with perfectly entrained steps. Across the road, the unimaginatively named "SHOE REPAIRS" shop brought up a memory of buying a pair of shockingly expensive bootlaces there a few years earlier when in some kind of urgent rush.

A woman approached me offering a small sprig of heather wrapped in tinfoil and the words "*heather, God bless you.*" I reflexively blurted out "*no thank you*" and turned away, feeling a slight jolt triggered by a memory from around 1992, having naïvely accepted such an offering from a woman in the High Street, then been subjected to some aggressive and unconvincing palmistry as she slipped into "gypsy fortune teller" mode and subsequently demanded £20 for her services (£10 for each hand). A deeply uncomfortable experience. My landlady at the time who worked in the shop Siesta said she'd once come in and angrily put a curse on the place.

A man with a blue charity bucket stood on the corner of the high street. Left turn. A young woman with a dog and retro 90's style "crusty" look (combat boots, etc.) passed. I'd seen her around the UKC campus for years – a postgrad? She'd recently cut off her dreadlocks. Right down Butchery Lane, unaware that I was passing Coca-Cola residues on the kerbstone "altar" which Yanik and his new acquaintance Tom had mock-blessed the night before.

11:53:48 to 11:55:03 on August 15th 2014

Passing the City Arms pub, its A-frame blackboard on the pavement featuring the city's coat-of-arms: three choughs (Becket's family heraldry) and a golden leopard in a posture known to heraldry enthusiasts as *passant gardant*. Two men and a woman were sitting outside enjoying late morning drinks. I turned to look at the admission price for the Roman Museum across the street: £8 for adults. I wondered if my local Residents' Card would get me in for free. A few steps further, Club Burrito. The idea of a burrito suddenly appealed. I'd been walking for hours. But as well as being on a mission, I reminded myself that I wasn't too well-off financially at the moment. This involuntarily triggered the execrable band Simply Red's 1985 single "Money's Too Tight (to Mention)" in my mind, which in turn produced a memory of a party in Whitstable in autumn 1988, the highlight of which was a drunken rant from a young working class man who'd torn this record off the stereo while a couple of middle-class drama students were dancing to it, shouting "*You're all living a lie!*" at the stunned revellers.

The numbers took me left down Burgate into the Buttermarket where tourists were milling around as usual. Suddenly I heard a familiar voice singing "*People, people, people, come on get on board...*"

11:55:03 to 11:56:04 on August 15th 2014

It was Josh Parnell, at that time a ubiquitous busker in Canterbury's streets, always playing reggae and wearing a Rastafarian "tam" hat. He was singing a slight variation on Bob Marley's "Zion Train" (1980):

"...get on board... Thank the Lord... Got to try to make it for station [?]*"*
[not, as Bob sang, *"I gotta catch a train 'cause there's no other station"*]

Someone standing nearby and clearly not paying attention was saying "*M&S and, and, um, Pret...*" I got out a pound coin. I'd often dropped Josh some change, appreciating the fact that someone was bringing righteous roots reggae to the city centre. This time I felt additionally glad that he was going to be documented in a history book and wanted somehow to connect with what he was doing.

"...then we're heading the same direction... yeah-I..."

I tossed my coin into his guitar case and got a perfectly timed "thank you". As well as his busking (which seemed to be as much about spreading a vibe and a message as making money) I'd twice seen Josh fronting the local band Hey Maggie: a triumphant gig at The Ballroom and then a slightly weird one at The Phoenix on the Old Dover Road, an odd choice of pub for a reggae band. I wandered on towards Sun Street, appreciating the steadiness and solidity of his "skank" strum as he morphed "Zion Train" into another, vaguely familiar, song.

11:56:04 to 11:56:52 on August 15th 2014

Josh's voice faded as I headed down Sun Street thinking about what a powerful presence he had as a singer. I paused to look at The Canteen, a takeaway whose falafel wraps I'd been recommended four years earlier (by Adam from the Furthur collective) but not yet experienced.

I continued on past a slightly worried looking woman in red trousers pushing a bike in the other direction, across from one of the many tacky souvenir shops aimed at daytripping tourists. I overheard a German voice saying "𝔍𝔢𝔡𝔢𝔯 𝔨𝔯𝔦𝔢𝔤𝔱 𝔫𝔬𝔠𝔥 𝔢𝔦𝔫 𝔄𝔫𝔡𝔢𝔫𝔨𝔢𝔫" which I later discovered, rather disappointingly, just means "*Everyone is getting a souvenir.*"

"*LAWRENCE!!*" shouted a woman at her little son in a strong local accent, then more calmly and mercifully: "*You be careful!*" The sounds of a creaking door and seagulls squawking were to be heard as I passed La Trappiste (a cafe/bar specialising in Belgian beers) and approached the beginning of Palace Street where a large white lorry was parked.

11:56:52 to 11:57:30 on August 15th 2014

"OMG!!" shouted a hand-chalked A-frame blackboard, "YOU'VE JUST WALKED PAST PORK & Co. THE BEST PULLED PORK ROLLS IN TOWN". A few paces further, Mrs. Jones' Kitchen announced via the same medium that its "Summer Specials" included iced coffees and "sparkling afternoon teas". Sparkling teas? Back in early May, the Free Range series of weekly avant-garde events had migrated from the Veg Box Cafe to Mrs. Jones' Kitchen, a much more suitable location with considerably more space and a windowed frontage onto Sun Street providing an ever-changing view of city centre life. I hoped this would last (it didn't). I wandered up Palace Street past the American Pancake House, a rather contrived Anglo-American cafe concept based around the fact that local puritan Robert Cushman had negotiated hire of the *Mayflower* there in 1620. Until fairly recently it had been "The Mayflower Restaurant & Pilgrims' Bar".

A father and daughter paused to glance into the window of Canterbury Glass Art while a fast-paced group of four lads walking in the street overtook me, one of them clutching an energy drink can, another protesting that "*'e's got a MASSIVE driveway!*" I passed the latest iteration of Dumbrell's, the tiny greengrocer run by three generations of the same family since 1948, previously located in St. Dunstan's, Lower Bridge Street and Military Road. The occupant of an approaching pushchair sipped from an orange plastic cup as I came up alongside master jewellery maker Ortwin Thyssen's shop and below another circular plaque for Mary Tourtel (this one marking the 1874 birthplace of Rupert Bear's creator).

11:57:30 to 11:58:00 on August 15th 2014

I crossed the street and turned to look down the curious narrow cul-de-sac called Turnagain Lane, high weaving lofts flanking its right-hand side, the only surviving example from the city's 17th century Walloon wool- and silk-weaving heyday.

Turning back to look up Palace Street, the four lads (Rick, Bill, Hugh and Reg, let's say) were walking directly in front of me, the one with the can nonchalantly disposing of it atop a stone pillar outside Conquest House, the medieval timber-framed building where Henry II's knights Reginald FitzUrse, William de Tracey, Hugh de Morville and Richard le Breton had plotted Thomas Becket's murder in late December 1170.

The pillar had been installed there a few years ago as part of the "King's Mile" development, helically inscribed with "*infinite riches in a little room*", a phrase from Christopher Marlowe's 1589 play *The Jew of Malta*. According to at least one less-than-scholarly source, this play was staged at the Guildhall in 1592 when Marlowe was visiting his family (all we know is that he visited Canterbury that year and that a play was staged, but not what it was). That was the same visit during which he'd fought a duel with a Cathedral musician at the Chequer of Hope Inn and ended up in court.

The route of Palace Street deviates from that of the Saxon *burgh* as Archbishop Lanfranc had it moved in order to build his palace in the late 11th century. I crossed back over to the Archbishop's Palace side of the road, wondering about the bricked-up archway in the wall.

11:58:00 to 11:58:25 on August 15th 2014

The numbers were steering me left down St. Alphege Lane where a pale yellow lorry was reversing with accompanying repetitive beeps. On the corner, a green felt hat sat upturned on the pavement as if placed there by a busker – but it was empty, this wasn't a busking spot and no busker was to be seen. "*That's Ziggy's hat!*" I immediately thought. Ziggy was the homeless Irish guitarist who I'd seen around a lot in the previous weeks, having first encountered him sitting in the parallelogram-shaped doorway of the "Crooked House" further up Palace Street while walking past with Miriam one July evening. He was burning incense and very drunk, but gentle and funny, babbling on to us about meeting guitar god John McLaughlin on a beach in France.

Leaning up against No. 6 was an old-fashioned bicycle with a basket. I'd often seen it there and assumed that it belonged to Orlando Bloom's mum Sonia who, I was told, lived in that house (she'd actually moved to Tyler Hill in 2009 but was still using it as an office address). Back in the 1930s it had been home to Richard Bellamy, leader of the Canterbury branch of Oswald Moseley's British Union of Fascists, serving as the local BUF headquarters and shop, openly selling fascist literature.

11:58:25 to 11:58:44 on August 15th 2014

Passing some scaffolding up against St. Peter's School of English (established 1966) on my right, the workmen who were drinking coffee and swearing at each other a few hours ago were still at work across the road, just outside the far corner of St. Alphege's churchyard. One of them eyed me suspiciously, possibly having remembered seeing a strange-looking person with a box under his arm mumbling into a microphone earlier in the morning. I felt a mild wave of paranoia that he'd spotted my hidden camera and hurried on towards the corner with King Street and Best Lane.

St. Alphege's Church had been deconsecrated in 1982, becoming the Canterbury Urban Studies Centre. It eventually became the Canterbury Environment Centre before being acquired by The King's School and closed to the public. I'd attended jazz gigs there as a student and last set foot inside in October 2011 when The Orchestra That Fell To Earth (former members of Penguin Cafe Orchestra) had played as part of the Canterbury Festival. It had been established by Archbishop Lanfranc in the 1070s and was the parish church of puritan Robert Cushman (who'd negotiated the hire of the *Mayflower*). Cushman was married there in July 1606.

Past No. 3, home of Cathedral lay clerk (choral singer) and insurance agent William Gough and his family in the 1880s, then No. 2, at that time home to a Mrs. Harriet Walker Bing of a local "ladies' school". She probably would have fainted if she'd been told that Canterbury would have a strip club bearing her surname a century-and-a-bit into the future. The drone of a light aircraft was audible overhead as I walked into a fairly strong breeze.

11:58:44 to 11:58:59 on August 15th 2014

The aircraft noise faded, along with the sounds of the workmen's tools, as I turned right down King Street. The now empty pub on the corner had been a short-lived gay bar called both "CO2" and "G-Bar" in 2009 after having been a failed lap-dancing venue called Scribes. Most recently, it'd had a short spell as Cromwell's Wine Bar. In the 1700s (when bull-baiting was still common) it had been the Dog and Bull, changing its name to The Prince of Wales in 1798 when the future King George IV had been presented with the Freedom of the City and sticking with that name until at least 1991.

A woman walking in the street with a bag on her left shoulder approached me as I walked alongside King Street Studios. This was The King's School's new photography and pottery centre, ceremonially opened in November 2012 by former pupil and award-winning ceramicist Edmund de Waal. His father Victor had been Dean of the Cathedral from '78 to '86 and, as described in his family memoir *The Hare with Amber Eyes*, was descended from the Ephrussi family, a Russian Jewish banking and oil dynasty. Almost exactly a year after I walked by, the website Culture24 had begun an article on de Waal: "*If Grayson Perry really is the people's potter, then Edmund de Waal is surely the thinking person's ceramicist.*" The good-humoured, iconoclastic and working-class Perry had first been enthusiastically brought to my attention by my artist friend Tom Langley, he of the rubbish raft regatta. I'd first met Tom in 2010 when he was doing lightshows for the Furthur collective. In contrast with the King's-School-and-Cambridge-educated de Waal, he'd grown up on the notoriously rough Sturry Road outside Canterbury. He and his girlfriend Sophie now lived a few doors further down King Street with their pet rabbit Mr. Nibbles. And yes, that's the same Sophie who got to look after Robert Wyatt on his visit back in July.

11:58:59 to 11:59:11.7 on August 15th 2014

As the woman with the bag passed me, children's voices were audible in the distance, blending with more light aircraft noise. Variegated ivy plants trailed from a pair of functional, hexagonal concrete planters outside Danecroft House. Most of the old houses along here were demolished in two phases in the first half of the 1960s, part of the slum clearance programme. Late 19th century residents had included Edward Miles (grocer), Frank Hills (gardener), Thomas Mason (general dealer) and Stephen Knott (painter).

The bollards along the edge of this pavement had been installed on 5th May 2005 to prevent, as far as possible, vehicles from mounting the kerb to pass and to park. At an East Kent Housing meeting that September opinions were divided as to their success. A resident of The Friars claimed that they'd resulted in noticeably slower traffic and reduced traffic flow, improving his/her enjoyment of this part of the city, whereas a King Street resident opined that heavy vehicles were causing danger to pedestrians and damage to pavements by trying to mount the kerb *between* the bollards.

A church bell began to ring. It didn't register with me at the time that this might mean the all-important 12:00 moment had arrived. I'd intentionally not brought a timepiece with me, but had set a phone alarm for 12:05 so that I'd know I could stop, then later figure out exactly where I'd been at 12:00:00 GMT via my time-coded audio and video. The random numbers were going to send me down Mill Lane. I crossed over towards the near corner, walking around a black Renault Mégane in a disabled parking bay.

11:59:11.7 to 11:59:21.5 on August 15th 2014

Turning left down Mill Lane, the street sign indicated a dead end (there's a row of bollards blocking vehicular traffic) and a small blue arrowed sign beneath it indicated that I was following a cycle route to "University / St Dunstans". I approached a nicely worn old brick wall with a hop plant growing over it from a King Street garden. Two student-aged males were approaching from some distance away, puffy cumulus clouds visible in the vivid blue sky above them. Walking with entrained steps, both wearing messenger bags, their body language suggested that they were deep in conversation.

According to census data, late 19th century residents of this end of Mill Lane included Mary Ann Robbins, her occupation being given as a glovemaker, and Bramwell Shufflebotham, a tattooist. Mr. Shufflebotham died in Canterbury in 1915 at the age of 44. Private Samuel Shufflebotham, son of Bramwell and his wife Emily, died in WWI in September 1918 and was buried in Zonnebeke, Belgium. Mary Ann Robbins died aged 70 in the tiny community of Exline, Iowa, having borne seven children.

11:59:21.5 to 11:59:29.3 on August 15th 2014

I suddenly remembered visiting Jacob at a house somewhere along here in early June 2010 to pick up some equipment needed for my 40th birthday party. He was an unusually deep-thinking philosophy student (seriously into Wittgenstein) and experimental musician who I'd occasionally see at local gigs and events, but that was the first time I'd encountered his soon-to-be-girlfriend Juliet (who I'd seen three times yesterday in the city centre, for the first time in almost a year). He'd been in a post-rock band called Bardo Thodol with various housemates and members of the UKC Psychedelics Society. Was it this house here on the right?

A black Mini was parked beside a large buddleia bush, just past the row of bollards which prevented the passage of traffic beyond the first few houses. The young men passed me on my right-hand side. No, Jake and co.'s house was further down.

11:59:29.3 to 11:59:35.6 on August 15th 2014

I passed between the two central bollards, at which point Danecroft House no longer blocked the direct line between the Cathedral bell tower and my ears, so louder, more authoritative "bongs" began to reach them, one about every 2.8 seconds.

I was following the northwest boundary of the medieval Blackfriars estate which would have been delineated by a wall to my left. The first piece of this estate, now Solly's Orchard, was granted to the Dominican brothers by Henry III in 1236-37. According to one historical source "...*the king permitted the friars to stop up a street leading to the mill of the abbot of St. Augustine in 1247, 'so that they made another road beyond a certain plot which the king had caused Stephen parson of Hadlinges to purchase with the royal money.'*" This road was Mill Lane, created by the Blackfriars in the mid-13th century.

11:59:35.6 to 11:59:40.5 on August 15th 2014

BONG!

Some fat white streaks were now visible in the sky and getting fatter. Rosy had first brought these to my attention when I was remarking on some unusual cloud formations a year or so earlier. She'd made a cryptic, offhand reference to "geoengineering" which led me into a headspinning labyrinth of online allegations, refutations and counter-refutations. Was some shadowy agency regularly spraying the Earth's atmosphere with toxic metal aerosols? And if so: (i) Who were they? (ii) What was their agenda? (iii) Wouldn't they and their families also be affected by the toxic effects? (iv) How were they keeping such a huge operation secret?

After delving into the website geoengineeringwatch.org, I was finding it hard to cycle around town without having my attention continually drawn to the skies (several near collisions had resulted). But I'd learned not to talk about this issue with most people as it tended to lead to heated discussions or accusations of mental instability. Miriam had also been looking into the geoengineering issue. The "pilgrimage" that she and I had made from Shepherdswell to Canterbury in March had been beneath a sky crisscrossed with these trails, being whipped up into cirrus-like cloud formations by the intense wind. I remember the goings-on in the sky being intensely distracting at the time and how we chose not to talk about this until later.

BONG!

11:59:40.5 to 11:59:44.5 on August 15th 2014

Ah yes, this was Jacob's house up on the right. Number 12. The unusually long house extended back to just touch a back corner of the 19th century Egyptomaniacal synagogue (now The King's School music recital hall) which faces onto King Street.

BONG!

I remembered him putting on a DVD of *loudQuietloud*, a documentary about the 2004 reunion of The Pixies, a band who'd been at their peak of musical creativity (and hipness) when I was a UKC undergraduate. Jake, Juliet, Oli and another friend were gathered around the screen in the smoky front room and I'd stuck around to watch part of it.

BONG!

11:59:44.5 to 11:59:47.6 on August 15th 2014

BONG!

To myself: "*...and I hear the church bells toll and I wonder...*" Was this 12:00, then? Best to just keep going and not dwell on the possibility.

Juliet had seemed rather away with the fairies that afternoon and didn't acknowledge me. I've since learned that she has no memory of this event and remembers first seeing me during the Sondryfolk arts trail ten months later, playing my saz on a swing in the Greyfriars Garden. Jacob had touchingly remarked on how happy the documentary made him feel. I hadn't heard the Pixies for years but immediately recognised the first song "Caribou" as being from their debut EP *Come On Pilgrim* (1987), which seemed to please him. It was strange to see an intelligent, arty, younger crowd getting into this music from my past and I felt a bit like I was time-travelling: arriving at Rutherford College in '88, the coolest students wearing Pixies T-shirts (Smiths T-shirts were on their way out). I'd followed the band quite closely and witnessed their decline but never saw them play live. The film was intriguing but some of their lyrical imagery now seemed unnecessarily brutal to me, which it hadn't back then. This induced a strong sense of time having passed and a sudden awareness of an incremental personal evolution.

I also remembered leaving with a mild feeling of disappointment that I'd not connected with Juliet. We'd become vaguely acquainted via Jake in the year that followed but then they'd moved away to Brighton. So I'd been pleasantly surprised to see her in town the day before.

11:59:47.6 to 11:59:50.1 on August 15th 2014

A particularly loud BONG! sounded as I reached the bend in the road.

A woman pushing a pushchair came into view, presumably having just passed the Millers Arms at the end of Mill Lane. I could hear her child begin to count: "*one...*"

Still thinking about Jacob and Juliet. Apart from just around the corner from here during that intense thunderstorm back in the autumn, the last time I could remember seeing her was after the Sunday night social event marking the end of Breaking Convention 2011, the first of the interdisciplinary psychedelic research conferences, held at UKC. He'd been at the next one in 2013, held in Greenwich, where he'd been on the "Psychedelic Music in Britain" panel that I'd been asked to chair, along with Nik Turner of Hawkwind fame, Twink from The Pink Fairies and former neighbourhood resident Steve Hillage's partner and synth pioneer Miquette Giraudy. Also on the panel was Anthony Saggers (a.k.a. Stray Ghost) who'd been in Bardo Thodol and also lived at no. 12. Jake had self-released an impressive album called *Through a Hole in the Head* as "Yakobfinga" in late 2012 after spending a difficult summer interning for eccentric aristocrat and trepanation enthusiast Countess Amanda Feilding's Beckley Foundation.

In the late 1800s, 11 Mill Lane had been home to the family of John Todd, a "general dealer". In November 1894 his young son Alfred and three friends were caught stealing from Jane Knott's sweetshop in Oaten Hill. For this they were each sentenced to twelve strokes of a birch rod by a constable, apart from Alfred who only had to receive six, being younger than the others.

11:59:50.1 to 11:59:52.1 on August 15th 2014

BONG!

Mother and rolling child approached. "*two...*" intoned the child. The bongs from the Cathedral were now less muffled, somehow more distinctive. Blue and red recycling crates were obliquely stacked on the pavement between nos. 10 and 11, part of the wildly unpopular "six bin" scheme introduced by the council in conjunction with bungling contractors Serco the previous summer. Blue for glass, tins, cartons and plastic and red for paper and card.

According to the 1901 Census, previous residents of 10 Mill Lane included a wheelwright named William Cobb and George Vile, a gas fitter and plumber. A John Vile, his wife Charlotte and their three sons had been living in Mill Lane back in 1841 so it's possible that George was one of the three (or perhaps a grandson).

At that time, a row of ancient cottages had occupied the other side of the road (to my left). These were selected for slum clearance in 1964, being declared unfit for human habitation. Conservationists had become a significant force by this point, though, so this didn't go without protest. City Architect John Berbiers drew up plans for some modernist houses to replace the cottages which were demolished in 1965-66 (around the time The Wilde Flowers were playing regularly at The Beehive in Dover Street). Berbiers' successor dropped those plans and was responsible for the large neo-Georgian houses at the King Street end of Mill Lane, disparagingly described by critics as a "miniature Chelsea".

11:59:52.1 to 11:59:53.7 on August 15th 2014

"*three...*" As she passed, I saw that the woman pushing the pushchair was texting.

Having spent the morning randomly walking around the city and looking carefully at what was going on around me, two of the most dominant impressions I was left with were: (1) mothers with children; (2) people absorbed in their smartphones. The former was a timeless phenomenon, something linking the Canterbury of August 2014 to the Cantwaraburgh and Durovernum Cantiacorum of old, and even earlier. The archetype was embodied by Our Lady Undercroft in the Cathedral Crypt and the Romano-British goddess Dea Nutrix. The latter impression, though, would have seemed entirely alien to almost everyone who'd previously walked these streets, something utterly of the early 21st century. This young mum, the last person I was to see before the moment which ends this story, perfectly combined the two.

"*four...*"

11:59:53.7 to 11:59:55.0 on August 15th 2014

BONG!

As they passed me, and I passed the recycling crates, I was struck by the strangeness of having not quite been provided with a countdown, but instead a "count-up".

The red front door of 3 Blackfriars Street and a pile of purple rubbish bags stacked up against a brick-and-flint wall appeared ahead of me. More hops were trailing over the slatted wooden fence on my left.

11:59:55.0 to 11:59:56.0 on August 15th 2014

Having reached "four", the child suddenly exclaimed "ooooh!", which at the time sounded to me like joy tempered with a sigh ("ohhhhh!").

11:59:56.0 to 11:59:56.8 on August 15th 2014

Listening back to my recording later, though, I could only hear the joy. I still have no idea what, if anything, s/he was counting.

11:59:56.8 to 11:59:57.5 on August 15th 2014

BONG!

To the Zoom H2 digital recorder I'd been carrying and speaking into all morning I dictated the words "*kid counting up to four and then sighing*" followed by an intake of breath.

In 1910 the red door directly ahead of me would have led to the home of 54-year-old widow Sarah Maria Graham and her children Ethel (24), Cecilia (20) and Sidney (16). That July, Francis Buckle and his wife Grace, to whom Sarah was renting an apartment, were convicted of stealing an antique jug from her, worth ten shillings. Grace confessed to selling on the jug but escaped punishment. Her husband was sentenced to a month's hard labour at HMP St. Augustine's.

Twenty to thirty years earlier, the house was inhabited by James Goulding (a cabinetmaker like his father) and his sister Mary, both born around 1818 in Sheldwich. They were part of an extended family of Gouldens in the Canterbury area, cabinetmakers and stationers, with a large shop at 39-40 High Street (opposite Guildhall Street – in 2014 it was a Vision Express and a Pizza Hut). A John James Goulden, born in Canterbury in 1841 and trained as a journeyman cabinetmaker, moved to Dover in 1865 to set up a stationery business. He and his wife Charlotte had four children including Richard Reginald Goulden (1876-1932), a celebrated sculptor responsible for several prominent WWI memorials.

11:59:57.5 to 11:59:57.99 on August 15th 2014

The mixed red brick and black flint wall against which the rubbish bags were piled was built directly into the foundations of the original Blackfriars precinct wall from the 13th century, the only remaining trace of that structure. It was now part of the garden wall for the house just around the corner on Blackfriars Street, a Grade II listed building (listed on 7th September 1973), described as a "'Gothic' flint cottage" in the October 2010 city council document "Canterbury Conservation Area Appraisal".

The house had most recently been bought for £140,000 on 10th September 2001, the day before the infamous "9/11" attacks in New York and Washington. At some point in the years since, an application had been made to the Canterbury City Council Planning Office to install an electrical meter box on the side of the dwelling (application no. CAL/02/00039).

Augmenting the hop plant (*Humulus lupulus*) and a climbing rose plant (not in flower) growing over the adjacent wooden fence, in some crevices on the wall's outer face were growing little ivy-leaved toadflax plants (*Cymbalaria muralis*).

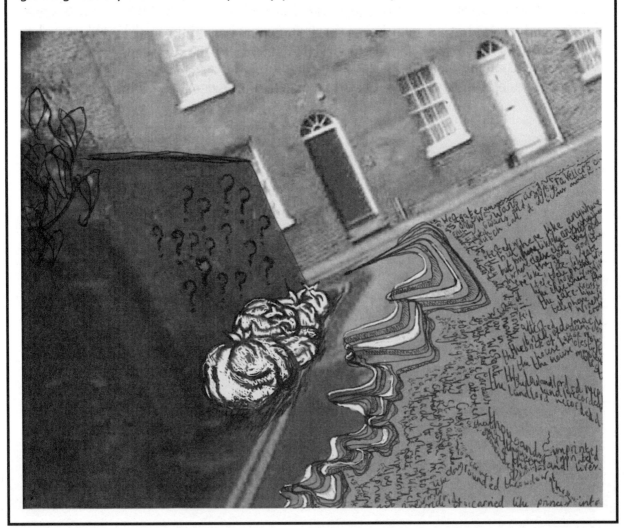

11:59:57.99 to 11:59:58.40 on August 15th 2014

To my right was 8 Mill Lane, originally two separate houses (nos. 8 and 9) which had been merged into one, now with two blue front doors opening out onto the street. In 1911, Blanche Grigg was born at no. 9 to Mary and Charles "Stoker" Grigg (first picture), as was her sister Grace in 1914. The house was home to Mary's brother Bill Weatheral (seated, second picture) whose children grew up there (including Fred, seen in the third picture with his bleached-out cousin Blanche). Charlie was well known in Canterbury. He'd been a stoker in the Royal Navy and was a formidable boxer, once winning the Army & Navy championships in Malta. When he was discharged in 1917 he opened an electrical goods shop at 35 Northgate (in 2014 this was Cut Above barbers), teaching boxing there as well as at the nearby Prince of Wales Youth Club. When the fair came to town, he would take on all comers over three rounds. An industrious type, he also made sweets and ice cream in a small factory behind the shop, Grace recalling how she, Blanche and their brother Perce were expected to help. He had a stall in Longmarket where he sold his produce, and an Indian motorcycle and sidecar from which he sold ice cream locally.

Charlie's father, Joseph Stockham Grigg (fifth picture), rode in the Charge of the Light Brigade at Balaklava in the ranks of the 4th Light Dragoons. He took his discharge in Canterbury in 1869 and settled, working as a warder in Canterbury Prison – the same prison that Grace's son Kevin, his great-grandson, worked in for twenty-five years until it was closed in early 2013. No. 8 Mill Lane was now owned, coincidentally, by a member of Kevin's local archery club (Canterbury Archers, founded in 1955).

11:59:58.40 to 11:59:58.72 on August 15th 2014

I must have cycled or walked the length of Mill Lane a couple of thousand times at least but I'd never given it much thought. Before Jacob had described where his house was I'd hardly even registered that it was called Mill Lane.

It was part of perhaps the most obvious route for cyclists and pedestrians coming down from UKC (across Beverley Meadow, down Station Road West, over The Causeway) to get into the city centre. This meant a steady stream of academics and students during term time. But Mill Lane also made up part of the most direct route from St. John's Coach Park, so daily hordes of tourists (including the dreaded French schoolkids) would walk from their coaches along the path beside the Stour to St. Radigund's Bridge, past the Millers Arms, down Mill Lane, to King Street, Orange Street, Sun Street and on to the Cathedral.

The other notable instances of human traffic I'd regularly seen along here were the controversial barrow traders who sold cheap souvenir crap to Euroteens from their wheeled carts in the high street. By the 1990s the density of them had become a major issue in the local press and had led long-term local MP Julian Brazier to introduce the "Canterbury City Council Bill" in Parliament in 2008 (passed 2013), locally "*altering the exemption enjoyed by holders of a pedlar's certificate from the street trading regime in the Local Government (Miscellaneous Provisions) Act 1982*". Since this became law they were fewer in number, but some were still to be seen coming down Mill Lane to and from St. Radigunds Car Park each morning and early evening, pushing their barrows stacked with baseball caps sporting zeitgeisty language fragments and cannabis leaves, "I ♥ LONDON" T-shirts, cheap sunglasses, Union Jack keyrings and tacky jewellery.

11:59:58.72 to 11:59:58.98 on August 15th 2014

As it had been 2.8 seconds before and 2.8 seconds before that, the three-tonne bell hanging in the Arundel (northwest) Tower of the Cathedral was struck with a hammer triggered by an electronic circuit. This was a descendant of a system installed in 1981 when Great Dunstan – as the bell had become known – was moved from the Oxford (southwest) Tower after having been cleaned and tuned at Whitechapel Bell Foundry in East London (Britain's oldest manufacturing company).

11:59:58.98 to 11:59:59.188 on August 15th 2014

The air pressure wave caused by the striking of Great Dunstan had travelled 376 feet, passing over the sundial in front of the Archbishop's Palace in the Cathedral Precincts, over the Buttermarket, over the mysterious upturned hat on the corner of St. Alphege Lane (if it was still there), over the Bell and Crown pub and out over Orlando Bloom's mum's former garden.

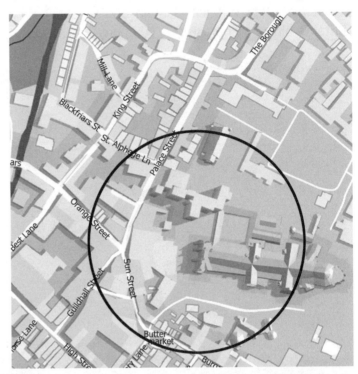

The bell continued to reverberate energetically. It had been recast from an earlier version of itself in October 1762 after a crack had developed from tolling with a hammer. This was done within the Precincts by master founders Lester and Pack of Whitechapel.

The version of Dunstan which had cracked in 1758 had (after a failed attempt of fix it with solder) been recast from a yet earlier version in 1684, during the later stages of Canterbury's silk-weaving boom, by London bell founder Christopher Hodson. It was declared to weigh 69.75 hundredweight (3543.46 kg).

11:59:59.188 to 11:59:59.353 on August 15th 2014

The soundwave emanating from the Arundel Tower had now travelled about 567 feet on the ground, over King Street Studio, reaching the "miniature Chelsea" at the King Street end of Mill Lane, the Beaney "House of Art & Knowledge", The Cuban bar, The King's School's Mint Yard Gate and the Thomas Becket pub in Best Lane.

Still Dunstan vibrated forcefully, continuing to produce a note close to a concert B. The bell from which it was recast in 1684 was itself recast from a bell in 1663 by Michael Darbey of Stepney. Weighing in at 62 hundredweight, the 1663 version had been seen as unsuccessful, hence its brief (for a cathedral bell) twenty-one-year tenure.

11:59:59.353 to 11:59:59.484 on August 15th 2014

The soundwave had now travelled about 717 feet, reaching Jacob and Juliet's old house at 12 Mill Lane. It simultaneously reached the former Blackfriars refectory (currently being restored) at the bend in Blackfriars Street, the old Post Office building (now a Prezzo restaurant), the Cherry Tree pub on White Horse Lane, Waterstones bookshop on St. Margaret's Street, Ossie's Fish Bar on The Borough (as endorsed by Archbishop Justin Welby) and the Old Synagogue.

The bell which Michael Darbey had recast had been hanging in the Oxford Tower since 1499 when it had been recast from an even earlier bell by John Bale. At that time, Canterbury was recovering from the bitter divisions between local factions of Yorkists and Lancastrians which had erupted during the Wars of the Roses and was still to enjoy almost another four prosperous decades of pilgrims flooding into the city before King Henry VIII ordered the destruction of Thomas Becket's shrine and ended the pilgrimages.

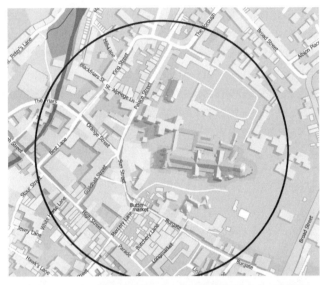

11:59:59.484 to 11:59:59.589 on August 15th 2014

The bell which John Bale had recast in 1499 had only been hanging in the Oxford Tower for forty years. It had been blessed and then installed in 1459 after a twenty-nine-year wait down in the Nave.

BO...

The spherical pressure wave broke on my eardrums, causing two sets of tiny bones to vibrate in sympathy with Great Dunstan, ultimately triggering electrochemical nerve impulses. Within a few dozen milliseconds, these had reached the auditory cortex of my brain and been processed, the reality of the ringing bell then somehow registering in my awareness. Almost simultaneously, the wave broke on the microphone of my handheld digital recorder, causing the modulation of an electric current (approximately 3 volts DC, produced by two AA batteries), this encoding the sound of the mighty bell along with other ambient sounds. The signal was passed through an analogue-to-digital conversion circuit, producing a set of integer outputs destined to be written to a 4 gigabyte SDHC memory card in MP3 format (256 kbps, 44000 Hz) five minutes later when I switched off the device.

My attention was being drawn to the pile of purple rubbish bags against the wall up ahead of me. Although the bags had been introduced by the city council over a year earlier, and despite thinking of myself as an observant person, this was the first time that I'd been consciously aware of their existence.

Light waves which had left the surface of the Sun around 11:51:40 (while I'd been walking down Rose Lane recalling cinematic traces of seventy-year-old sunlight) plunged through the Earth's atmosphere in just over a millisecond and reflected off the bags towards my eyes. The quantum chemistry of the molecules in the plastic which were responsible for its distinctive purple colour caused the reflected light to have a vibrational frequency somewhere in the range of 6.68 to 7.89×10^{14} Hz, the waves arriving at my retinas via my pupils just fractions of a microsecond later.

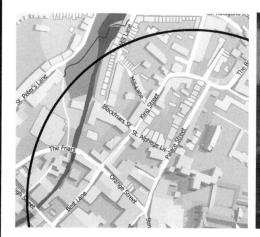 

11:59:59.589 to 11:59:59.672 on August 15th 2014

Great Dunstan is recorded as having been given to the Cathedral by Prior Molash in 1430, having been cast in London (by an unknown bell founder) and weighing 72.5 hundredweight.

...ON...

The soundwave had now travelled about 930 feet along the ground and extended along the path of the city wall almost exactly from Skunkworks and the The Jolly Sailor (next to the site

of the old Northgate) along Broad Street, past The Victorian Fireplace and the Canterbury Archaeological Trust office, to the site of the old Burgate, then round past the taxi rank on Canterbury Lane, Tiny Tim's Tearoom on St. Margaret's Street, Canterbury Wholefoods on Jewry Lane, Third Eye in St. Peter's Street, the far side of the Marlowe Theatre, Solly's Orchard and Hawkswell's picture framing shop on St. Radigund's Street.

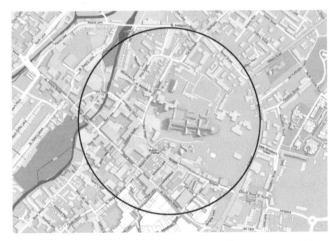

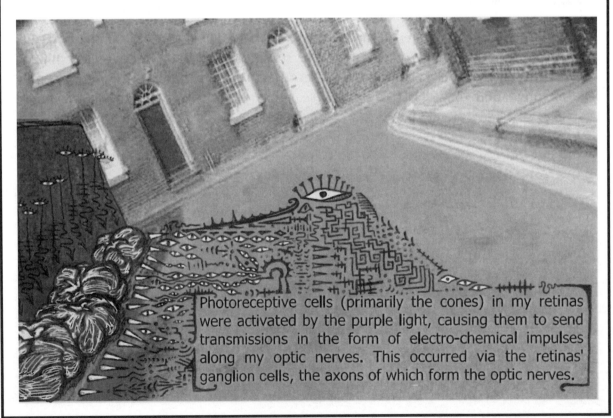

Photoreceptive cells (primarily the cones) in my retinas were activated by the purple light, causing them to send transmissions in the form of electro-chemical impulses along my optic nerves. This occurred via the retinas' ganglion cells, the axons of which form the optic nerves.

11:59:59.672 to 11:59:59.739 on August 15th 2014

The bell was intended to replace an earlier bell called "Dunstan" which had hung in the campanile, gifted to the Cathedral in 1343, along with a bell called "Jesus", by Prior Hathbrand. Dunstan and Jesus were destroyed when the campanile fell in the earthquake of 1382. All of this went on while Canterbury was experiencing the peak of its pilgrimage period.

...NG!

The soundwave had now travelled 1006 feet on the ground, reaching as far as the remains of the Abbot's Mill at the end of Mill Lane, the Oxfam bookshop on St. Peter's Street, the northeast tip of Binnewith Island in the Greyfriars estate, the Bagpuss Museum on Stour Street, the Primark in Whitefriars, the clocktower on St. George's Street, the Magistrates' Court on Broad Street and "Transylvanian", a specialist Romanian food shop on Northgate.

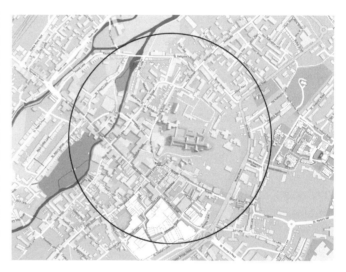

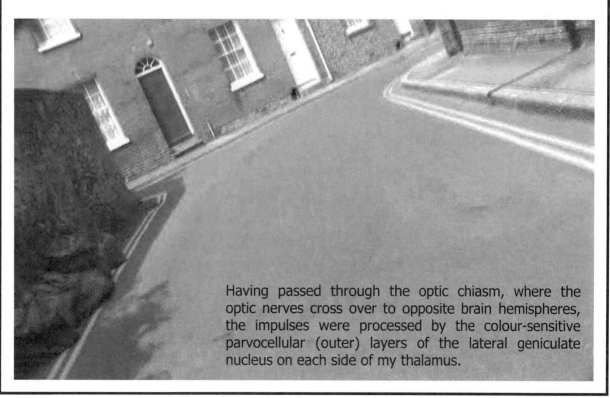

Having passed through the optic chiasm, where the optic nerves cross over to opposite brain hemispheres, the impulses were processed by the colour-sensitive parvocellular (outer) layers of the lateral geniculate nucleus on each side of my thalamus.

11:59:59.739 to 11:59:59.792 on August 15th 2014

At this point the soundwave had travelled 1066 feet on the ground in all directions out from the Arundel Tower, reaching as far as the Hobgoblin music shop on Lower Bridge Street, Zoar Chapel (strict and particular), Wilko (née Wilkinson) in St. George's Street, the former Slatters Hotel on St. Margaret's Street (ten years derelict), the Poor Priests' Hospital on Stour Street (now the Heritage Museum), St. Peter's Methodist Church, the Holistic Health Clinic on Northgate and the statues of King Ethelbert and Queen Bertha on Lady Wootton's Green.

The purple binbags were also linked to the widely despised "six bin" scheme introduced in spring 2013 when outsourcing company Serco won the contract for refuse collection in the district. From then on, rather than going to the former landfill site near Broad Oak, the purple bags of household waste were transported to a site in Sandwich, anything non-recyclable then being transported to a plant in Allington (near Maidstone) and incinerated for energy production.

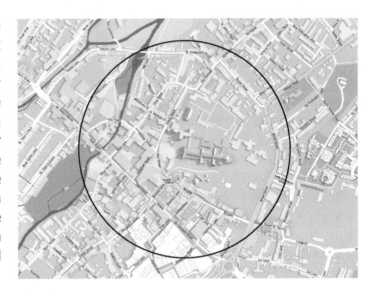

The superior colliculus in my midbrain was now activated and began to (via my oculomotor nerves, back through the optic chiasm) use my extraocular muscles to direct my eye movements around the pile of bags.

11:59:59.792 to 11:59:59.834 on August 15th 2014

The soundwave had expanded as the surface of a sphere, now of radius 1114 feet, having reached as far as the site of St. George's Gate, St. Paul's Without the Walls (near the Abbey's "Cemetery" or "Ethelbert" Gate), The Lady Luck in St. Peter's Street, Westgate Hall (with its handsome new red-finned ventilation cowls) and Starry Oriental Grocery on Northgate.

On each of the purple low-density polyethylene bags was printed the Canterbury City Council logo, featuring a silhouetted view of a simplified cathedral and a stylised tree. Bell Harry, the central tower of the depicted cathedral, loomed up behind me. Earlier that morning in the very same building, a confusing passage had been solemnly read aloud from an old storybook:

[10] Moreover the Lord spake again unto Ahaz, saying, [11] ask thee a sign of the Lord thy God; ask it either in the depth, or in the height above. [12] But Ahaz said, I will not ask, neither will I tempt the Lord. [13] And he said, Hear ye now, O house of David; Is it a small thing for you to weary men, but will ye weary my God also? [14] Therefore the Lord himself shall give you a sign; Behold, a virgin shall conceive, and bear a son, and shall call his name Immanuel. Butter and honey shall he eat, that he may know to refuse the evil, and choose the good.

Ahaz had been a king of Judah in the 8th century BCE. In 2014 the majority of his former kingdom was part of the Occupied Palestinian Territories, at this time very much in the news, as a tense ceasefire was just about holding in Israel's relentless bombardment of the Gaza Strip ("Operation Protective Edge"). In 2012 human rights activists in East Kent concerned with Israel's 45-year military occupation of Palestinian lands had protested vigorously to prevent the French multinational Veolia from being awarded the city council's new refuse collection contract. Veolia was involved in refuse collection and transport infrastructure serving Jewish-only settlements in the West Bank (where, curiously, many of the colonists justified their blatant violations of international law by citing the same old storybook). It appears that the council had attempted to award them the contract through the backdoor but were eventually forced to hold a ballot. The efforts of the "Bin Veolia" campaign (backed by the Bishop of Dover, among others) may have been a major factor in the contract being instead awarded to Serco.

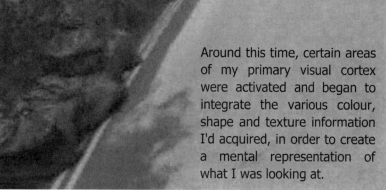

Around this time, certain areas of my primary visual cortex were activated and began to integrate the various colour, shape and texture information I'd acquired, in order to create a mental representation of what I was looking at.

11:59:59.834 to 11:59:59.868 on August 15th 2014

Dunstan's voice had now travelled 1152 feet along the ground, reaching the Abbey's Fyndon Gate (early 14th century), the Chaucer Bookshop on Beer Cart Lane (est. 1956) and The Causeway.

In February my old friend Sarah F had moved into a house in St. Dunstan's, just opposite the church. She and her friend Megan were working at Eurostar in Ashford and had jointly decided that it would be a lot more fun to live in Canterbury, a place they regularly visited. I asked her about this some time later:

me: "...and when you moved in, you hadn't heard of the purple rubbish bags?"
Sarah: "I'd NEVER heard about the purple rubbish bags." [laughter]
me: "I think they'd been in the national press at least twice by then..."
Sarah: "I must have missed that particular scoop!" [more laughter]
me: "So what was your first encounter with them?"
Sarah: "Well, we put our rubbish out... Obviously we moved in, we had loads of rubbish, we didn't know when the rubbish dates were... we missed the first rubbish date. And then the week after we were like, right, we'd better sort this out, let's put all the rubbish out. So we put it in black bags, came back in the evening and our rubbish was still there in front of our doorstep."
me: "Right."
Sarah: "And, um, we rang up Serco to say..."
me: "So you knew Serco were responsible?"
Sarah: "Yeah, you can just go on the Canterbury Council website and find out. So we rang them up and said 'What have we misunderstood? Why is our rubbish still with us?' And they

said 'Was it in a purple bag?' [much giggling]
To which we said 'of COURSE it wasn't! It was in a BLACK bag!' [more giggling]
'Ah! That'll be the problem!' they said, 'You can come down and get some of 'em from us.' But luckily our neighbour had a surfeit of purple bags and she gave us a couple of rolls, and our rubbish was collected the next week."
me: "OK. And you've not had any trouble since?"
Sarah: "No, absolutely, it's fine. As long as it's purple, it's good."

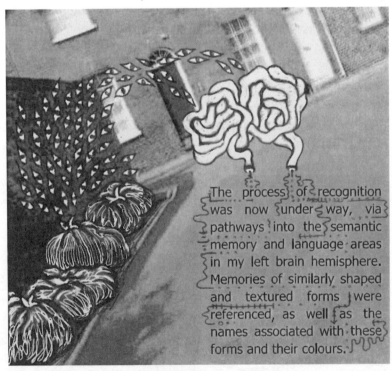

The process of recognition was now under way, via pathways into the semantic memory and language areas in my left brain hemisphere. Memories of similarly shaped and textured forms were referenced, as well as the names associated with these forms and their colours.

11:59:59.868 to 11:59:59.895 on August 15th 2014

The pressure wave now extended 1183 feet in all directions on the ground from the point below Great Dunstan, having reached Funky Monks and Revivals in St. Peter's Street, the playground of St. Thomas Catholic Primary School on Military Road and Super Noodles on Northgate.

Downstairs at the Roman Museum I'd passed a few minutes earlier on Butchery Lane is a beautifully preserved floor mosaic from a Roman townhouse dating to around 300CE. It's about two metres below the current street level. Canterbury's street level rose an average of just over 1mm per year until relatively recently due to the great volume of physical material brought in compared to the amount being taken out: building materials, food scraps, bones, broken pots, bodies... Roads had been periodically maintained by adding new layers of gravel (quarried from nearby pits in the Stour valley), contributing to this process, but a lot of the total volume beneath the rising city would have been some form of rubbish.

In more recent years this rising tendency had ceased with people's conveniently bagged rubbish (these days polystyrene packaging, sponges, paper towels, food scraps, tampons...) being collected as part of a scheme organised by the council. Until the new scheme was introduced in 2013 it had been transported to the Shelford Landfill site near Broad Oak, about a mile and a half northeast of the city centre. When I returned to Canterbury in 2007 and was working on the top floor of UKC's Templeman Library, I could see a nearby hill just north of the Downs Road housing estate. I wondered why I'd never noticed it before, why I'd never explored it during my student years. I soon found out that this was because *it wasn't there then*, it effectively being a gargantuan pile of household waste, landscaped, complete with methane vents and occasionally roved over by maintenance vehicles. So, rather than the city centre rising due to an endless net influx of physical material, an area of comparable size and shape a couple of miles away had been rising in its stead.

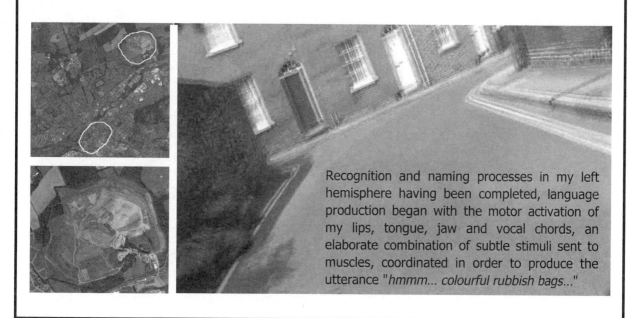

Recognition and naming processes in my left hemisphere having been completed, language production began with the motor activation of my lips, tongue, jaw and vocal chords, an elaborate combination of subtle stimuli sent to muscles, coordinated in order to produce the utterance *"hmmm... colourful rubbish bags..."*

11:59:59.895 to 11:59:59.916 on August 15th 2014

The outer edge of the soundwave had now achieved a radius of 1207 feet, reaching St. Thomas Catholic Primary School itself, Kenny Hunter's *Lamb* statue near the bus station, The Three Tuns pub (a site once centre stage of the 3rd century Roman amphitheatre), Water Lane Cafe and the Sidney Cooper Gallery on St. Peter's Street.

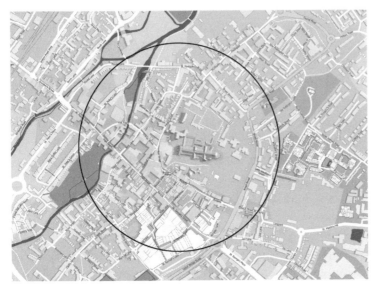

The banal observational utterance *"hmmm... colourful rubbish bags..."* formulated a few dozen milliseconds earlier – my intention being to commit it to my stream-of-consciousness sound recording – began to emerge from my throat as a vibration with an approximate frequency of 300 Hz (in other words, 300 oscillations of my vocal chords each second).

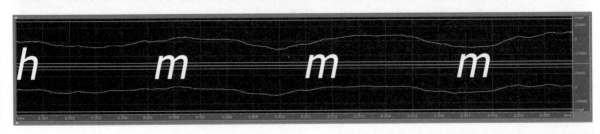

11:59:59.916 to 11:59:59.933 on August 15th 2014

At 300Hz (approximately) the eye at the centre of the spiral twitches under the lid. If it opens, it will meet the quizzical gaze of the professor, staring down at it through a magnifying glass and looking for purchase on its slippery terrain: a quicksand ever winding inwards, the infinite in the infinitesimal.

Or turn the hourglass up the other way. Somewhere in the later recurrent sixes of the 59.91666...th second of the 59th minute of the hour, the barrier between worlds loses resolution, fragments, and all perspectives collide and fuse together. At the inner constrictions of the spiral's centre, things run very tight, and there is not room for the multiple points of reference entertained in the earlier and more expansive coils. Instead, all must be as one: small is big and big is small; what is momentary is also eternal. The detective becomes the detainee, and the one-eyed man wakes up to discover he is king.

So up-end the hourglass, and observe from that inverted angle. Observe the ant, his runes cast, his course plotted and followed: stood now outside the temple at the centre of the labyrinth, rearing on his hind legs to face the sleeping giant he has tracked to her lair. Observe the arid maze of time and space that stretches without limit in all directions from him. You are here: the ant draws his circle on the map, the game is begun, and there are not a few local spirits who will take an interest in these peculiar pastimes enjoyed by people, ants and ghosts.

The ant professor has done well! He has led a merry dance and amassed quite an entourage to parade with him through the streets of Canterbury in an increasingly frantic spiral-search for the eternity in the heart of an instant.

11:59:59.933 to 11:59:59.947 on August 15th 2014

But now the sleeping eye twitches, and the crowd judders to a stop. Great Dunstan's hammer is poised, but does not fall: his dome tenses. The excited stream of Vikings flowing out of the cathedral is suddenly plugged, and the Parliamentarian snipers who had begun to swarm down over the gate freeze. Invicta grinds to a halt across the Bullstake. The Archbishops on board have all fallen to their knees, and one is hurriedly pulling off his Guy Fawkes mask. Wild pigs jostle to a bumbling stillness, and Julian Brazier silences a gaggle of anxious petitioners with a glare. The bugle falls flat with an embarrassed squeak; there are those who will swear that before the banner crumpled in the slackened grip of awestruck hands, the red lion's tail was between its legs. A veritable horde of avant-garde musicians have lain their instruments reverently at their feet. At Lady Wootton's Green, Bertha holds her breath and Ethelbert, swung in the direction of the enchantment, finds himself spellbound; Hlothhere and Mul too are paralysed. The wind sneaks down through the Quaker gardens (the grass does not stir, nor the leaves on the trees) and darts below the bridge to hide with the bats and two huddling brides in the damp shadows. Dea Nutrix has locked eyes with Mother Mary, as though their steady gaze could stabilise the already cracking silence.

A witch titters, tied to the ducking stool. Above the king's meadow, a crow bestows a ragged word upon the stillness. At 12 Mill Lane - where moments before, a Pixies documentary had flickered to black, where in the downstairs front room a cello named Sylvia had stifled the last resonance of a low yawn and now lay silent - a black three-legged cat jumps down from the wall. With the air of one in search of something lost, he meanders over to the frozen ant professor and sniffs the fallen magnifying glass at his feet. Closer inspection reveals it is in fact a neolithic jadeite axe. Marcel rubs his furry cheek on its smooth face and purrs on back towards no.12.

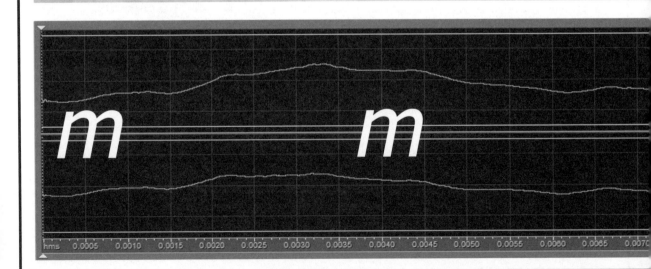

A little way around the corner there is a square courtyard adjoining the Egyptian synagogue gardens. It is walled in by black brick and sand-coloured stone. The gate is black, the leaves and briars gold. In the very middle of the square is a tree about five feet tall, pruned to a perfect globe on a thin stalk. There is a girl standing by the tree, dressed for winter. She is peering into the tangled, thorny centre of the sphere. Suddenly, she steps back. Seeming not to sense the otherworldly tension that holds all else in its thrall, she skips across the courtyard, and stops to smile at a little, round, cloaked and hooded friar hunched furtively in the corner. He looks at her as though holding back a terrified moan. Proferring a quick bow, she slips out of the gate; more cautious now, she hurries towards Mill Lane, flanked by red brick housefronts, studiously ignoring the row of street merchants clustered on the road ahead of her. Rounding the corner onto Mill Lane, she stops to stroke Marcel. At last, she acknowledges the absolutely still gathering of ghosts upon ghosts, and sits back on her heels, observing them in startled amusement for some time before she shrugs and gets up to go. Her eye is caught by a gold, fox-shaped knocker on the door of no.11.

She lifts it.

She drops it.

CLUNK

The eye rolls beneath its lid.

The girl slips into the crowd and disappears.

The eye opens.

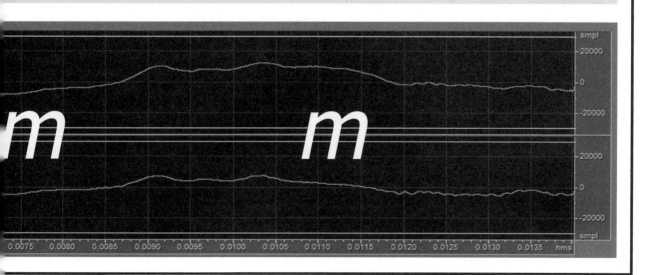

After a staring contest that seems like it will never end, the giant drops the ant's gaze and looks at her toes. She sniffs loudly and wipes her nose with her fist.

WHAT DO YOU WANT?

"I - I'm very sorry, but... who are you?" squeaks the ant. The giant stares again.

IN THIS PLACE I AM BECOME, AND WHO KNOWS TO WHAT EXTENT THE GENIUS LOC! DWELLS IN ME?

Several ghosts shuffle their feet. Tom Holden, crouched amid the toadflax and lavender, thumbs his acoustic bass guitar.

HERE TIME KEEPS HER SECRET COILS. HERE, AS IN EVERY OTHER HERE IN TIME AND SPACE, EACH IS A WELL-LAID TRAP: AN ATTENTIVE EYE IS LIABLE TO BE DRAWN EVER DEEPER INTO THE LABYRINTH. THE MORTAL WHO SETS HIS COORDINATES TO THE BEAT OF THE MYTHIC DRUM RANDOM INVITES A SPIRIT WHOSE FORM IS COY, BUT WHOSE METHOD IS NOT.

The ant frowns. "I'm sorry, would you mind elaborating on that last bit?"

BLUNTLY PUT, MORTAL: YOU ARE ASKING FOR TROUBLE!

The ant gulps. "I'm not! No trouble!"
The giant laughs.

TRUBBA NOT, DETECTIVE. THE SEEKER WHO REMAINS THE COURSE WITH PURE HEART WILL BE LED WELL ASTRAY, BUT NEVER TO PERIL, AND ON THE SOLSTICE NIGHT, A DARK TRAIPSE IN FIELDS UNKNOWN IS INEVITABLE, A THICKET IS A TRANSFORMATION, AND THE SUNRISE IS A TREASURE. THE ARCHETYPES APPLAUD, THE PALACE WALLS RESOUND!

"So... I won? This was supposed to happen - I did well?"
The giant narrows her eye.

THE SPIRAL CHARTED WITH THE HEROIC INTENT OF THE HUNTER-EXPLORER APPRENTICING HIMSELF TO A GRAVE QUEST. THE COIL ISOLATED, THE SNAIL KIDNAPPED AND HELD TO RANSOM. TIME BURGLED! TIME PERPLEXED! SO I RETURN TO MY ORIGINAL QUESTION: WHAT DO YOU WANT?

The ant boggles. "I don't mean to be rude, but you aren't being very consistent, sir - madam; one minute, using the word idiomatically of course, you tell me I've pleased you, the next that I've robbed you, and it wasn't my intention to do either! I don't want anything, I just want... to..." he falters.
The giant tilts her amorphous head.

MY GOOD PROFESSOR. I AM AN ARTIST. I AM NOT A BROKER OF DEALS. I AM NOT A FAIR MAN, AND NO ONE HAS EVER CALLED ME DIPLOMATIC. BUT I RECOGNISE YOUR VERY VALID COMPLAINT. YOUR PETITION HAS NOT FALLEN ON DEAF EARS. THE PROBLEM IS, THERE DOESN'T SEEM TO BE ANYTHING HERE. AND YOU'RE RIGHT. THERE SHOULD BE. I SHOULD HAVE SOMETHING I CAN OFFER YOU. I MUST HAVE SOMETHING I CAN GIVE YOU.

The ant simply stares in rapt bafflement at this absurdity, this abject paradox, that he has awoken at the heart of time.

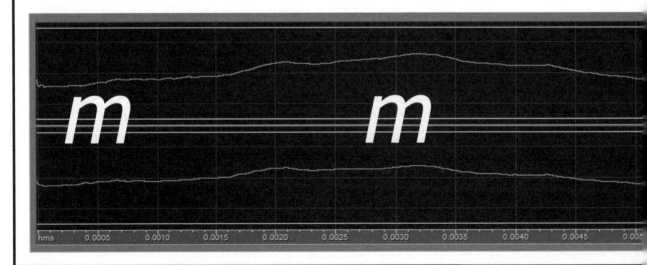

I THINK I KNOW THE SORT OF THING YOU'RE AFTER. YOU'D LIKE A NICE LITTLE ETERNAL MYSTERY AND THE FLEETING SENSE OF HAVING UNDERSTOOD ITS SUBLIME CONSTELLATION. WOULDN'T YOU?

The giant looks patiently at the ant. The ant blinks back, dumbstruck – then inhales deeply and gathers himself to himself. "Well... do you have anything less... abstract?"

LESS ABSTRACT! O HO, NOW HE KNOWS WHAT HE WANTS! HOW ABOUT THE NOTE, THE SINGLE NOTE SINGING IN THE AIR, TO WHICH ALL OTHER NOTES ARE DRAWN AND RESOLVE? YES, I SEE THAT APPEALS TO YOUR EARNEST NATURE, YOUR POETIC HEART. OR A GLOWING GOLDEN GLOBE, A SINGULAR THING OF OTHERWORLDLY BEAUTY?

"Ohhh," breathes the ant. "Like... the Aleph? Like in that story by..."

I OFFER HIM CONCERT B, AND HE ASKS ME FOR THE ALEPH! THE ALEPH, INDEED. THOSE ARE FAR MORE TROUBLE THAN THEY'RE WORTH. IF YOU ARE LOOKING FOR THE JEWEL IN THE HEART OF THE LOTUS, LOOK AROUND YOU! THE HIDDEN HELIX AND THE SINGLE SINGING NOTE, THE ALEPH AND THE GLOWING GOLDEN GLOBE – THEY LIE BETWEEN EACH OF ITS PETALS, YOU DAFT BEAST! DO YOU NOT KNOW THAT IF YOU GO SEARCHING FOR THE TRUE ARTICHOKE, YOU WILL END UP DIVESTING IT OF ITS LEAVES?

The ant is resigned. "I am sorry. I did not know." The giant looks surprised, and then smiles.

NO. I AM SORRY. I MUST LEARN TO CONTROL THESE TEMPERS. YOU ARE RIGHT TO SEEK SOMETHING THAT EXCEEDS THE EVERYDAY WONDERS OF CREATION. AND I KNOW NOW WHAT I HAVE TO GIVE YOU, MY DEAR, BRAVE, VALIANT PROFESSOR.

The giant produces a bag and holds it out to the ant.

CRAFTED IN THE FINEST, INDESTRUCTIBLE SILK, THE MILK OF THE TREE IN THE HEART OF THE EARTH, THE WOOD IN THE HEART OF THE STONE. SUCH A BEAUTIFUL, VIBRANT SHADE OF PURPLE. AS THOUGH IT WERE THE FIRST SHED SKIN OF AN AMETHYST. SEE HOW IT SHINES IN THE NOONDAY SUN! GO ON, MY LITTLE SHINING MAN. TAKE IT. IT'S YOURS.

The ant takes the bag.

OPEN IT, GO ON. YOUR PRIZE, YOUR LONG SOUGHT CENTRE LIES WITHIN...

The giant can barely contain herself. The ant professor opens the bag.

MMMMMM... COLOURFUL RUBBISH BAGS WE HAVE HERE, AREN'T THEY?!

...and with a cackle that ripples to the outermost reaches of the spiral, she is gone. The snake swallows its tail; the coil is sucked into its centre and vanishes.

GNAIS GIB

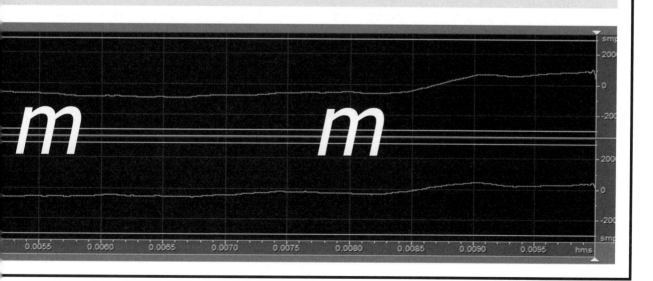

11:59:59.957 to...

1 Planck length = 0.0000000000000000000000000000000000161619... metres

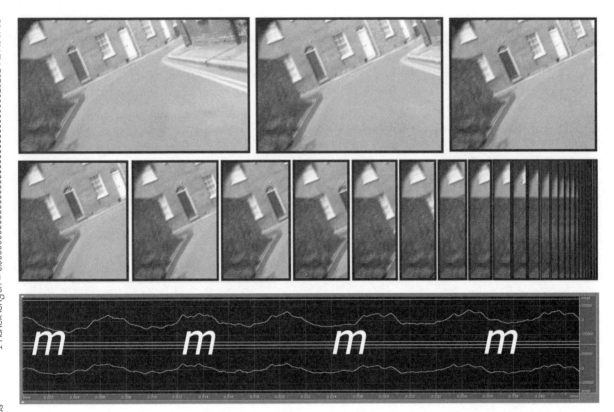

...12:00 on August 15th 2014

1 Planck time unit = 0.00000000000000000000000000000000000000000005391963... seconds

To be true to its underlying concept, this book would require infinitely many of the "episode frames" it's been built from: it doesn't matter how many times you multiply a quantity by 0.79696..., it will keep getting smaller but never reach zero, just as a pure geometric spiral will wind around its centre infinitely many times, getting ever closer but never arriving.

Whether time is infinitely divisible, though, is another question. After 420 of the frames shown above, we would be dealing with episodes covering less than $5.3919632 \times 10^{-44}$ seconds, the so-called *Planck time*. This is the time it would take a hypothetical particle moving at the speed of light to travel one *Planck length*. A Planck length is about 1.616199×10^{-35} metres, the distance below which conventional physics breaks down and quantum effects take over, the smallest distance it's possible to meaningfully measure. So, very roughly put, in less than one Planck time unit it's impossible for anything to happen... at least anything measurable. So, after that point, there really would be *nothing* to report in any single frame. And yet, confusingly, our eventful lives are built from a succession of countless oodles of Planck time units.

In any case, to create a visual impression of this infinite sequence of episode frames, I've stacked together the appropriate fractions of the final still image which my camera recorded in the last 0.042495 of a second before 12 o'clock.

In that final 0.042495 (slightly less than 1/23) of a second, the sound of Great Dunstan would have travelled a further 48 feet, arriving at the Masonic Temple in St. Peter's Street, the butterfly garden beside the Stour, St. John's Hospital (claimed to contain the "oldest toilet in England"), The New Inn pub in Havelock Street, the Cathedral side of the Abbey grounds, the music practice rooms in Christchurch University's Coleridge House (main campus), the bus station, Concorde House (a healing centre in Stour Street) and erstwhile gay pub The Carpenters Arms.

My vocal chords continued vibrating to create the "*mmm...*" sound emerging from my mouth, but the rate of vibration dropped significantly from approximately 300 Hz towards 250 Hz where it was to terminate at 12:00:00.635 when I would blurt out the words "*colourful rubbish bags*". This seems to have been an unconscious attempt to entrain my voice with Dunstan's (his last BONG! still ringing audibly down Mill Lane at about 248 Hz). So, in that 0.042495 of a second, my vocal chords would have performed only about a dozen oscillations, each taking fractionally longer than the previous one.

Also in that 0.042495 of a second, made up of about 800 million billion trillion trillion Planck time units, as well as cats sleeping, dogs barking, butterflies sipping nectar in front gardens, bees buzzing, rats gnawing, woodlice scurrying, worms burrowing and bacterial cells dividing in unknown numbers, something on the order of 5,000,000,000,000,000 neurons were firing in approximately 50,000 human brains in and around the City of Canterbury, their owners variously daydreaming, conversing, eating, sleeping, making love, making tea, making lunch, suckling babies, texting, reading, writing, flirting, scheming, watching cat videos on YouTube, composing letters in Microsoft Word, drinking beer outside pubs, rolling joints in bedrooms, washing, walking their dogs, shopping, praying, studying, sightseeing, busking, meditating, painting, decorating, driving, cycling, walking, jogging, skateboarding, crawling, tripping, watching the 1976 comedy film *The Gumball Rally* on BBC2, listening to DJ Fearne Cotton on Radio 1, scrolling through Facebook, browsing on eBay, swiping on Tinder, playing cricket, watching cricket, knitting, policing the streets, moving furniture, picking blackberries in marginal wild places, gardening on allotments, researching, debating, speaking the Queen's English, Estuary English, broken English, French, German, Polish and Romanian, arguing, apologising, listening, ignoring, advising, kissing, cuddling, hugging, laughing, crying, struggling to breathe, doodling, solving equations, looking in mirrors, taking selfies, waiting on tables, working on building sites, practising musical instruments, sneezing, hiccupping, burping, farting, pissing, shitting, sitting bored in meetings, looking at pornography, buying, selling, whispering, shouting, playing with toys, pretending, counting, scribbling, trying to remember something, trying to forget something, complaining, worrying, enthusing, whistling, getting dressed, getting undressed, working on crossword puzzles, playing board games, playing computer games, brushing teeth, flushing toilets, stressing out, chilling out, falling in love, falling out of love, switching on and off lights, opening and closing doors, preparing for the birth of a baby, reeling from the death of a parent, feeling toothache, feeling heartache, feeling overjoyed, feeling overwhelmed, getting angry, showing compassion, facing death and embracing life.

the aftermath

The air temperature was about 18ºC, a few degrees above the average for that time of year, the atmospheric pressure was a fairly average 1015 mbar and there were light winds out of the NNW. The Moon and Uranus were in Aries, Neptune in Pisces, Pluto in Capricorn, Saturn and Mars in Scorpio, and the Sun, Mercury, Venus and Jupiter were all in Leo.

My alarm went off at 12:05 and with a noticeable release of psychic pressure I switched off my audio and video recording devices and headed straight to the tiny shopfront "Anglican Catholic" church of St. Augustine in Best Lane to furtively observe their Assumption mass.

The door was wedged open and a sign on the pavement urged the pedestrian public to come in and join the worship. I chose to hang around outside, watching Bishop Damien Mead, together with some kind of deacon or right-hand man, an organist and a congregation of three. There was a lot of ritualistic incense censer swinging, elaborate costume-wearing and ceaseless reconfiguring of ecclesiastical paraphernalia during the

Anglican Catholic Church
St Augustine of Canterbury
Best Lane, Canterbury, Kent

Friday 15th August 2014
**Solemnity of the Assumption
of the Blessed Virgin Mary**

12 noon Solemn Pontifical Sung Mass
Celebrant & Preacher: The Bishop

Followed by
**Veneration of the Relic of the Veil of the
Blessed Virgin Mary**

Refreshments afterwards at
Conquest House, 17 Palace Street,

*Hail Mary, full of grace, the Lord is with thee.
Blessed art thou among women and blessed is the fruit of thy womb, Jesus.
Holy Mary, Mother of God, pray for us sinners now and at the hour of our death. Amen.*

ceremony but also a surprisingly accessible sermon from the Bishop in which he explained that the Assumption was essentially religious folklore as the Bible doesn't mention Mary after her son's resurrection. The circumstances of her death are unclear. But Bishop Mead argued that the lack of any "first-class relics" (physical remains) in circulation was good evidence that she had been magically absorbed into Heaven. We were urged to be more like Mary, answering the calling of the Lord as she had.

The main event, after the rather dull service, involved the veneration of a tiny piece of what was claimed to be the Virgin Mary's veil. The Bishop explained that this wasn't a "veil" in the current face-covering sense but rather a sort of undergarment. He told a story about one of Charlemagne's nephews or grandsons acquiring it and giving it to a French cathedral, and how it later got torn to bits by angry peasants in some kind of uprising. Monks had carefully gathered up the pieces and a "teeny-weeny little piece" (he smiled) had found its way to this tiny church in Canterbury. He stressed that they weren't going to be *worshipping* it but rather *venerating* it. I could have gone in and queued up with the faithful to kiss it – I was momentarily tempted – but I was feeling a bit paranoid with my hidden camera and spun out from the intensity of documenting the last few hours. Instead, I headed over to the Goods Shed, the farmers market where a few friends worked, to unwind with a drink at the Wild Goose bar and a chat with Lucy, its ever-cheerful proprietress.

I've since found out that the reliquary in that little church contains supposed first-class relics of St. Agnes, St. Cornelius, St. Godelina, St. Leocadia, St. Beggue, St. Anastasia, St. Anacletus, St. Simplicius and St. Donatus. As well as the acid-tab-sized piece of Mary's underwear, there are second-class relics from St. Thérèse of Lisieux and St. Margaret Mary Alacoque ("*All are authentically documented for veneration by the faithful.*").

The church had been opened in April 2007 with a blessing from Bishop Rommie Starks, a native of Covington, Kentucky. A bit of background reading revealed that this overlooked little place of worship is in fact an American export, the acting cathedral of the "Diocese of the United Kingdom" for a worldwide church based in the unlikely location of Athens, Georgia (best known for having spawned the bands REM and the B-52s in the 1980s). This emerged from the 1977 Congress of St. Louis where the American "Episcopalian" branch of the Anglican communion was divided over reforms including the ordination of women priests. Bishop Mead is an Englishman, a helpful front as he is effectively being presented as the "real [Arch]bishop of Canterbury". The Anglican Catholic church considers the former shop "*reclaimed by us from secular use and dedicated to our patron, St. Augustine*" to be the diocesan cathedral (the diocese being the whole UK), as it houses the Bishop's "cathedra" or throne. The "real Canterbury Cathedral", then. Also, an icon of Elizabeth Barton, the 16th century "Holy Maid of Kent", was commissioned by the church in 2008 (the first in the world) and blessed by Rommie Starks. They consider her to be a martyr.

Anglican Catholic Church of St Augustine, Canterbury, Kent updated their cover photo.
15 August 2014 · 🌐

Assumption Day 2014
The Relic of the Veil of the Blessed Virgin Mary under the statue of Our Lady in Church.

The little church on Best Lane was funded by a church supplies shop run out of Conquest House on Palace Street, which explains how it was able to operate with such a tiny congregation (I'd passed by on previous occasions to see only two or three worshippers in attendance). Considering the infamous role of Conquest House in the martyrdom of Becket, this siting of their shop struck me as odd, even provocative.

The Anglican Catholic Church seems to retrospectively agree with Henry VIII's break from Rome. They don't want to be affiliated with the Roman Catholic Church.

👍 Like   💬 Comment   ➤ Share

🔘 3

They just felt that the existing Anglican Communion wasn't "keeping it real", so they had to break away to maintain some true tradition. Did they also agree with Henry's retrospective stripping Becket of his sainthood and declaring him a traitor, destroying his shrine and abolishing the pilgrimages there, I wondered? Their Canterbury "cathedral" was dedicated to St. Augustine, not St. Thomas. But then again, they were now venerating relics, while their rival neighbour the "Cathedral and Metropolitical Church of Christ at Canterbury" had long since shunned such blatantly magical practices.

I came to realise that the city has been "running on magic" for the majority of its history. According to former Archbishop Rowan Williams, the distinguished Anglican scholar Dom Gregory Dix compared the early Saxon archbishops to "tribal wizards". They continued to mediate between Earth and Heaven as other types of magical practitioners before them had. The first ten Archbishops of Canterbury had been made saints with cults arising around their shrines. When Vikings were threatening to attack in the 10th century, the primary concern of the monks at the Abbey was not to conceal the treasure that might be looted but to dig up and hide the bones of the various deceased primates. The bones were where the real power was to be found. A particular type of magical belief found around the planet anchors a world of ancestors into an Earthly matrix of bones which are carefully preserved and venerated. Long before Becket, cults had sprung up around St. Dunstan (the most beloved saint of the English prior to Becket) and the martyred archbishop St. Alphege. Their shrines had attracted pilgrimages. And then, with Becket's martyrdom, 350 years of mega-pilgrimage began.

It took me a while to realise that all those pilgrims weren't just coming to Canterbury, to enter the city gates. Nor were they coming merely to enter the Cathedral. No, the idea was *to get as close as possible to Becket's remains*, which happened to be in a reliquary in a shrine in a cathedral in Canterbury. It was widely believed that proximity to the remains of his physical body could result in some kind of blessing or spiritual purification. This sort of thing is wholly rooted in a magical worldview and has nothing at all to do with the message of Christ. So we see a survival from pre-Christian tribal Saxon polytheistic and earlier animistic peoples. The early Norman archbishops did their best to suppress the Saxon Cult of Saints. It seems that the Saxons in Kent had adopted Christianity but then just matched up its various saints with their minor gods and goddesses of old.

In the months leading up to my chosen moment, I'd become increasingly convinced that a Marianist cult had been suppressed in the Canterbury area. It seemed strikingly relevant that at its medieval peak, about a third of Canterbury's numerous churches were dedicated to Mary, including chapels over several city gates and in the Poor Priests' Hospital. The same trend was apparent with the dedications of many nearby village churches. From figurines unearthed in the area we know that there was a pre-Christian mother goddess cult active in the area before Augustine arrived. I imagined that when presented with the Jesus story by his entourage, the Cantwaraburgh locals would have wanted to know "where's the Mother?" The only significant mother figure was Mary, so St. Mary became a new face of the Mother Goddess. The fact that she seems to have been by far the most popular saint in the area suggests that the pre-Christian motherhood cult may have been the dominant one. But if there *was* any such suppression, the Church had done it very quietly and effectively I had to admit, historical traces being well hidden.

The magical-Christian fusion continued on through the Saxon and Norman eras, finally being suppressed by Henry VIII with the Reformation in the 1530s, Becket's shrine looted and destroyed, his bones (allegedly) burned and the pilgrimages banned. But regular rituals continued in the Cathedral involving the consumption of Jesus' flesh and blood, the central ritual of Christianity. The Cathedral, like all the great cathedrals of Europe, was built as a structure in which to conduct this ritual. All the hymns, sermons, candles and stained glass were secondary diversions. Canterbury Cathedral was a place in which magic continued to be practiced.

After the Civil War, Canterbury fell into decline and became a minor provincial market town with an oversized church, the archbishops living in Lambeth, many not even bothering to visit. At the same time, the Church of England was starting to lose its grip on the people (a steady process which had led to an almost entirely disinterested local population by 2014), its "magical" or superstitious aspects in particular beginning to seem like a leftover of an earlier age. Methodism, Baptism and various nonconformist Protestant sects began to spring up, and eventually deism, agnosticism and atheism. Around the same time, though, we see the emergence of antiquarianism, a new kind of interest in the past – in ruins, artefacts, various remains of earlier ages. Wealthier members of English society with time for such things would visit places of antiquarian interest and Canterbury was among the most prominent.

This trickle slowly grew into the torrent of tourism we see now with coach parties, the Channel Tunnel, Eurostar trains and cheap air travel. The reasons people are coming to Canterbury keep changing and are often vague: it's very old... it's in the title of a very old book everyone's heard of but almost no one has read... stories about pilgrims... a great big, beautiful cathedral... cobbled streets and half-timbered medieval inns... So they flock here and have their photograph taken in front of the Cathedral or the Westgate, trying to partake of it somehow. They might be Australian backpackers touring Europe, American retirees who've watched too many BBC period dramas or French schoolkids on day trips who are supposedly being educated about history or culture or something. Or they might be Japanese or Italian Soft Machine obsessives.

Even if we factor out geographical considerations (being between London and the Channel), there seems to be something pulling these people, a belief that *there's something here* and that by coming here you can somehow get some of it. The name resonates with people, rings old bells. I was pulled here in 1988 when planning my return to England from a period in Wisconsin, trying to choose a university by looking at their printed prospectuses (no websites back then). The picture on the cover of the UKC prospectus – looking down on the city and Cathedral from the campus in a hazy light – drew me in and immediately made me think "*I could see myself there.*" Another factor was an awareness, via a small paragraph I'd seen in an encyclopedia of popular music, of a "Canterbury Scene" in the 60s and 70s. It include a list of bands I had no way of listening to at the time, but the idea that such a small city could have produced this outpouring of exploratory music made me think that the place might have some sort of special atmosphere about it.

This all makes me think of Hilaire Belloc pilgrimaging in from Winchester in the 1920s, looking for some ancient spirit that lingers but finding it in a tavern rather than in the Cathedral. Of Ed, Will and Ginger, former Cathedral choristers periodically pilgrimaging outward from Canterbury in the 2000s, but always being pulled back. Of Riddley Walker, the eponymous character of Russell Hoban's remarkable post-apocalyptic novel set in East Kent, who talked about the "Cambry Pul", pulling everything to the centre.

We could also consider the appropriation of the idea of pilgrimage by local language schools, the appropriation of Chaucer and Marlowe by local businesses and the council, the appropriation of Robert Wyatt and Caravan by UKC and CCCU, the appropriation of *Bagpuss* and *The Clangers* by everyone... The Anglican Catholics made a point of establishing their church (the "real Canterbury Cathedral") in sight of the Cathedral, dedicating it to St. Augustine, trying to tap into some of that power. The cover of original Caravan bassist Richard Sinclair's 1994 album *RSVP* shows him with the Cathedral visible in the background, a device he's used in other promotional photos over the years. Likewise, young local bands have been known to emphasise their geographical Canterbury roots in the hope of getting some interest from music writers vaguely familiar with the Canterbury Scene of yore.

In the 60s, UKC had its Rutherford and Eliot Colleges built oriented so that the main windows of their central dining halls faced directly onto the Cathedral. The Stephen Cox sculpture *Hymn* it commissioned in 1991 and the hillside labyrinth commissioned in 2008 were similarly oriented. UKC holds its graduation ceremonies in the Cathedral, these being extremely popular with students and their families, particularly the wealthy foreign ones. But, unlike the supposedly Anglican CCCU, there's no institutional link between UKC and the Church. It's an illusion, but a powerful backdrop against which to frame one's achievements, to create the impression that "this has got something to do with *me*". The Cathedral is lit up at night with powerful sodium lights, presumably at quite some expense, but it's not for the Glory of God these days, rather to generate a romantic nighttime atmosphere that will keep the tourists coming in.

* * *

On reflection, I realised that documenting a moment in the way that I did meant that the "content" of the moment was inevitably going to involve *me documenting something*. But I liked that, as (in my mind, at least) its self-referentiality seemed to resonate with the endless inwinding of spirals and the "Marianist recursion" I'd considered back in the spring when I was considering this date as an endpoint: Mary, revered as Mother Goddess of all Creation, being absorbed, or "Assumed", back into the matrix she herself embodies.

Einstein put forward the concept of "space-time" as a framework for the material Universe, where space and time are conjoined into a four-dimensional continuum, the points or locations within which are unique "events", each event an exact moment at an exact position in space. My aim was to capture not just the *genius loci* or "spirit of the place" in Canterbury but rather its space-time equivalent, what I'd call the "*genius loci-tempi*", the spirit of the place *at a particular time*. By its very nature, this is a fleeting entity, so attempting to capture it ended up being a curious exercise involving precisely time-synched technology, random numbers, "mythogeography", social media and far too many tedious hours in front of the Beaney's microfilm machine scouring back issues of the *Kentish Gazette*.

And if you argue, "yes, but you only captured the '*genius loci-tempi*' of the bend in Mill Lane at 12:00GMT on 15th August 2014", then consider this. According to "many-worlds" interpretation of quantum mechanics, whenever a situation has a number of possible outcomes, *they all happen*, in slightly different versions of the Universe, branching away from each other, leading to an unimaginably vast continuum of parallel universes at any given time (sometimes called the "Multiverse"). This idea was explored poetically by the visionary Argentinian polymath Jorge Luis Borges in his 1941 short story "The Garden of Forking Paths".

With this worldview in mind, I realised a couple of years after the event that every time I'd used the random number generator function on my calculator, a number of forks had occurred in the multiversal time stream, one for each outcome. Thinking along these lines, for every single location on the streets of the walled city, there would have been universes where I ended up there, as *any* possible string of random digits numbers would show up in *some* universe. Beyond that, there would have been anomalous universes where I would have been invited into strangers' houses through complex chains of events or been gripped by the urge to climb over a wall into someone's garden, or undertake some other form of trespassing. So there was an infinite ensemble of mes across all possible Canterburys at 12:00:00 that day, observing in careful detail what was going on around me, then researching the background to that moment and creating a version of this book. The book you're currently reading, then, could be seen as just one volume on an infinite bookshelf (spanning the Multiverse) capturing the *genius loci* of Canterbury at noon on the Feast of the Solemnity of the Assumption of the Blessed Virgin Mary, the Year of Our Lord 2014.

* * *

P.S. If you hadn't already deduced it, the extraordinary illustration sequence and handwritten narrative at the end were contributed by Juliet who, based on some random numbers, was the last person in my thoughts prior to 12:00, as well as someone I'd seen thrice the day before (for the first time in ages). After another period living in Brighton, she resurfaced at a Lapis Lazuli gig at Bramley's about a year later, in the late summer 2015. After I featured in an intensely precognitive dream of hers, we began a dialogue and soon became good friends, discovering a shared enthusiasm for Borges, Wittgenstein, Russell Hoban's Riddley Walker, Leonora Carrington, Alice Coltrane, Godspeed You! Black Emperor, psychedelia, Jung and shamanism. When I cautiously broke the news to her that she was going to feature quite significantly in a book I was writing, she was reassuringly enthusiastic about the concept and more than happy to get involved, with her consciousness starting to leak into the narrative at the point when I wandered past 12 Mill Lane, evoking goings-on there four summers earlier (and foreshadowing this with a few of her dream-drawings).

gazeteer

Note: There are ten maps on the pages that follow, numbered 0-9. The map references below work as follows: "**7**:(8.3,1.2)" means "See map 7. The place you seek is 8.3 units across and 1.2 units down from the upper-left corner." For streets, a pair of terminal locations is given. Quite a few street names have changed over the centuries.

Abbot's Mill **5-8**:(4.3,2.1)
Adisham **2**:(6.7,5.3)
Albion Place **9**:(6.1,2.8)-(6.3,2.4)
Alcroft Grange **3**:(6.7,1.1)
Aldington **2**:(4.7,8.6)
All Saints Church **7-8**:(3.7,4.4)
Allington **2**:(0.7,4.4)
Alma Street **9**:(7.2,0.7)-(7.4,1.0)
amphitheatre, Roman **4-5**:(3.7,6.4)
Archbishop's Palace **5-9**:(4.7,3.6)
Archbishop's School **3**:(6.3,2.3)
Army recruitment centre **9**:(3.0,3.3)
Ashford **2**:(4.3,7.4)
Augustine House [CCCU library] **9**:(4.6,8.5)
Avebury **1**:(5.2,5.4)
Aylesham **2**:(6.8,5.7)
Ballroom, The **9**:(4.4,3.9)
Barham **2**:(6.4,6.3)
Barming **2**:(0.5,5.3)
Barretts car showroom **9**:(2.8,3.2)
Basilica, Roman **4**:*location unknown*
Beaconsfield Road **3**:(6.3,3.2)-(6.4,2.6)
Beaney Institute, The **9**:(3.8,4.6)
Beehive, The **9**:(5.7,7.6)
Beer Cart Lane **5-9**:(3.3,5.7)-(3.7,6.4)
Bekesbourne **3**:(9.6,6.4)
Best Lane **5-9**:(3.7,4.4)-(4.2,3.8)
Beverley Meadow **3**:(6.4,2.9)
Beverley Meadow tunnel **3**:(6.5,3.2)
Beverlie, Ye Olde **3**:(6.5,2.6)
Bigbury Hillfort **3**:(3.5,4.7)
Bing, The **9**:(5.7,7.7)
Binnewith Island **5-9**:(3.1,5.4)
Bishopsbourne **3**:(9.1,9.9)
Black Griffin Lane **9**:(2.2,5.2)-(2.9,3.4)
Blackfriars dormitory **7-9**:(3.8,3.2)
Blackfriars Gate **7-8**:(3.4,3.9)
Blackfriars refectory **7-9**:(4.2,3.3)
Blackfriars Street **9**:(4.3,2.6)-(4.4,3.4)
Blean **3**:(4.7,0.7)
Blean Woods **3**:(3.6,1.4)
Bloomington-Normal, Illinois **0**:(1.6,2.4)
Boho Cafe **9**:(3.7,4.5)
Bossenden Woods **3**:(2.1,1.8)
Boughton-Under-Blean **3**:(0.7,2.4)
Bramley's **9**:(4.3,3.8)

Brittany **1**:(3.9,9.9)
Broad Oak **3**:(7.8,0.3)
Broad Oak Road **3**:(6.6,3.2)-(7.6,1.8)
Broad Street **5-9**:(5.5,2.2)-(6.0,5.8)
Broomfield **2**:(6.4,3.1)
Bullstake **5-7**:(4.7,5.0)
Burgate **4-8**:(5.9,5.7)
Burgate [Street] **5-9**:(4.7,5.0)-(6.0,5.8)
Burgate Lane **5-9**:(5.6,6.7)-(5.8,5.7)
bus station **9**:(5.2,7.4)
Butchery Lane **5-9**:(4.6,5.6)-(4.9,5.1)
Butterfly Garden **9**:(3.4,1.9)
Buttermarket, The **8-9**:(4.7,5.0)
Caen **1**:(5.7,9.2)
Cafe du Soleil **9**:(3.2,2.4)
Calais **1**:(7.3,6.6)
Cambridge **1**:(6.5,4.7)
Canterbury **1**:(6.9,6.2)
Canterbury Archaeological Trust offices **9**:(5.7,2.8)
Canterbury City Cemetery **3**:(5.7,3.3)
Canterbury East station **9**:(2.9,9.3)
Canterbury Heritage Museum **9**:(3.4,5.4)
Canterbury Lane **5-9**:(5.3,6.3)-(5.6,5.6)
Canterbury Open Centre **9**:(3.5,9.3)
Canterbury Tales, The [tourist attraction] **9**:(4.3,5.5)
Canterbury West station **9**:(2.5,0.6)
Canterbury Wholefoods [2000-2014 era] **9**:(3.7,5.4)
Caseys **9**:(4.7,5.3)
Castle, Norman **6-9**:(2.3,8.3)
Castle Street **6-9**:(2.4,8.5)-(3.7,6.4)
Catching Lives **9**:(3.5,9.3)
Cathedral **5-9**:(5.3,4.3)
cattle market **5-8**:(5.5,7.7)
Causeway, The **6-9**:(3.6,1.0)-(3.8,1.9)
CCCU main campus **9**:(7.7,4.4)
Chartham **3**:(3.6,6.8)
Chaucer Fields **3**:(5.9,2.6)
Chaucer School [Tech] **3**:(7.5,4.6)
Chequer of Hope Inn **7-8**:(4.6,5.5)
Chantry Lane **6-9**:(6.3,8.7)-(7.4,6.3)
Cherry Tree, The **9**:(3.7,5.1)
Chilham **3**:(1.1,8.3)
Christchurch Gate **7-9**:(4.6,4.9)
Christchurch priory **5-7**:(5.4,3.9)
Christchurch University [main campus] **9**:(7.7,4.4)
Church Lane **9**:(5.0,2,0)-(5.3,2.3)

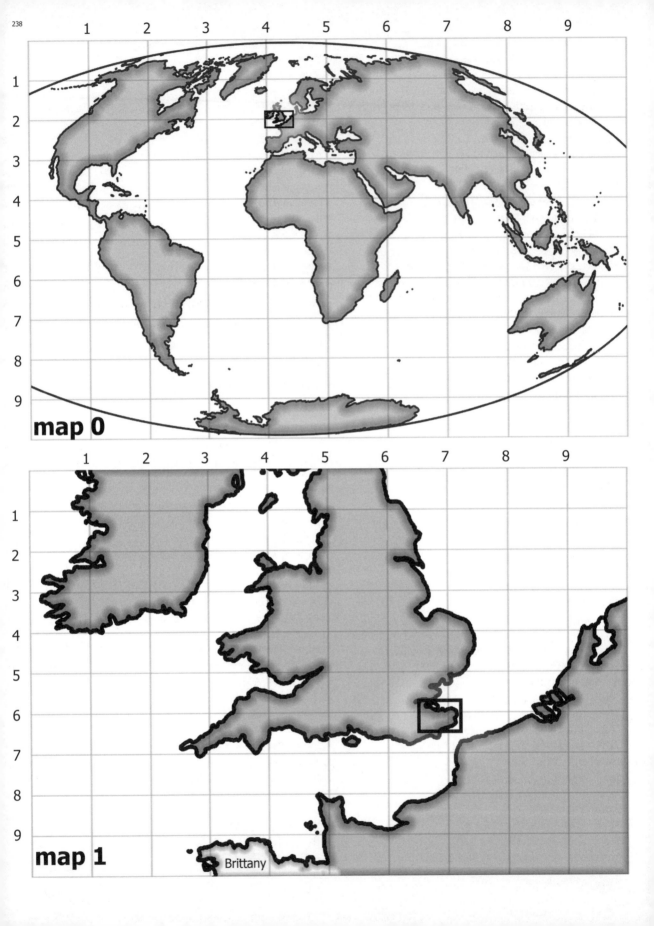

map 0

map 1

Brittany

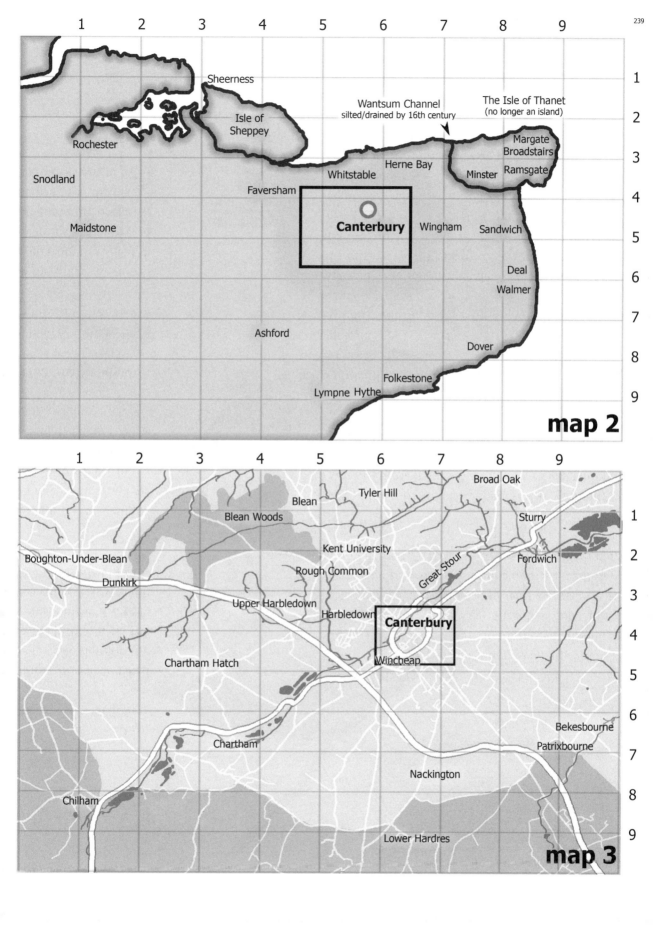

map 2

map 3

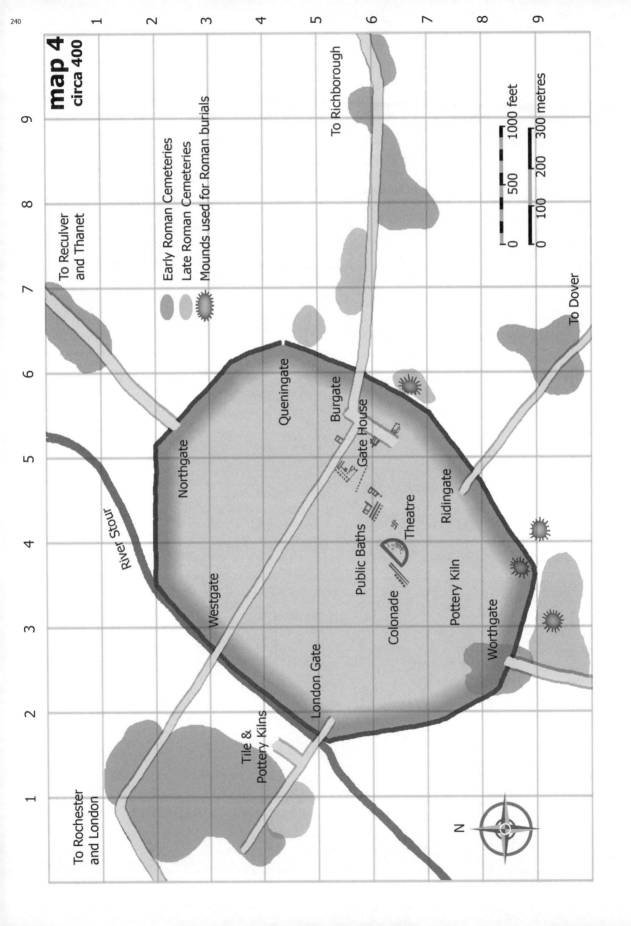

map 4
circa 400

To Reculver
and Thanet

Early Roman Cemeteries
Late Roman Cemeteries
Mounds used for Roman burials

To Richborough

To Dover

1000 feet
500
0
300 metres
200
100
0

River Stour

Northgate

Queningate

Burgate

Gate House

Westgate

Public Baths

Colonade

Theatre

Ridingate

Pottery Kiln

Worthgate

London Gate

Tile &
Pottery Kilns

To Rochester
and London

N

240

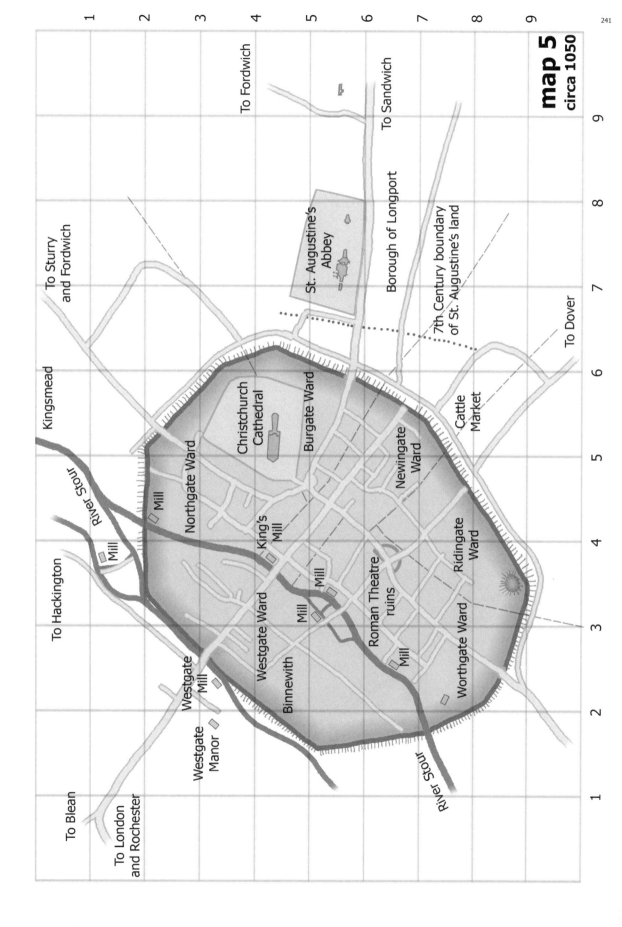

map 5
circa 1050

241

To Fordwich

To Sandwich

St. Augustine's Abbey

Borough of Longport

7th Century boundary
of St. Augustine's land

To Dover

To Sturry
and Fordwich

Kingsmead

River Stout

To Hackington

Christchurch Cathedral

Burgate Ward

Newingate Ward

Cattle Market

Northgate Ward

Mill

Mill

King's Mill

Mill

Mill

Mill

Westgate Ward

Binnewith

Roman Theatre ruins

Worthgate Ward

Ridingate Ward

Mill

Westgate Mill

Westgate Manor

River Stout

To Blean

To London
and Rochester

1
2
3
4
5
6
7
8
9

1 2 3 4 5 6 7 8 9

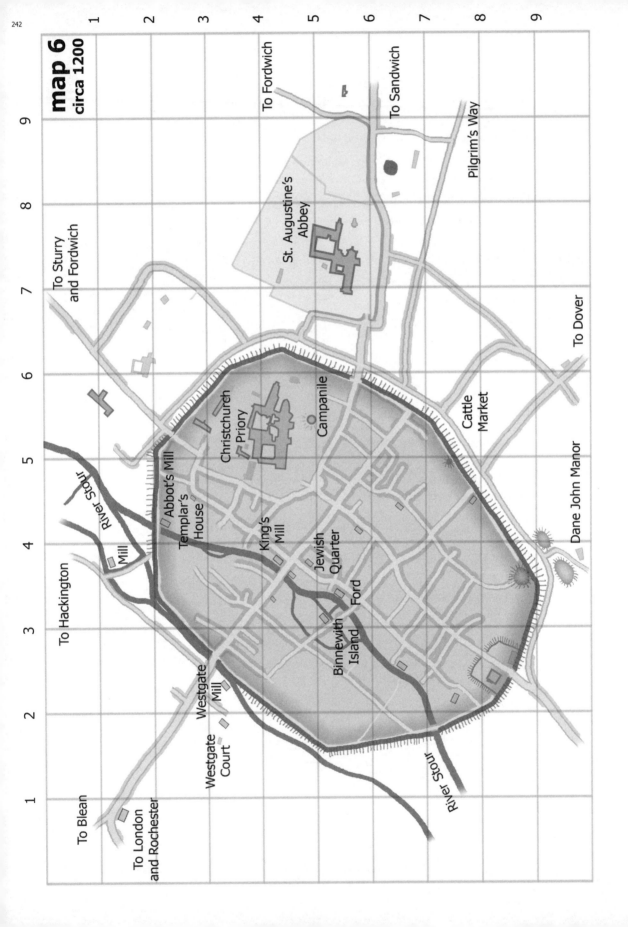

map 6
circa 1200

To Fordwich

To Sandwich

Pilgrim's Way

St. Augustine's Abbey

To Sturry
and Fordwich

To Dover

Christchurch
Priory

Campanile

Cattle
Market

Abbot's Mill

Templar's
House

River Stour

King's
Mill

Jewish
Quarter

Dane John Manor

Mill

Ford

To Hackington

Binnewith
Island

Westgate
Mill

Westgate
Court

To Blean

To London
and Rochester

River Stour

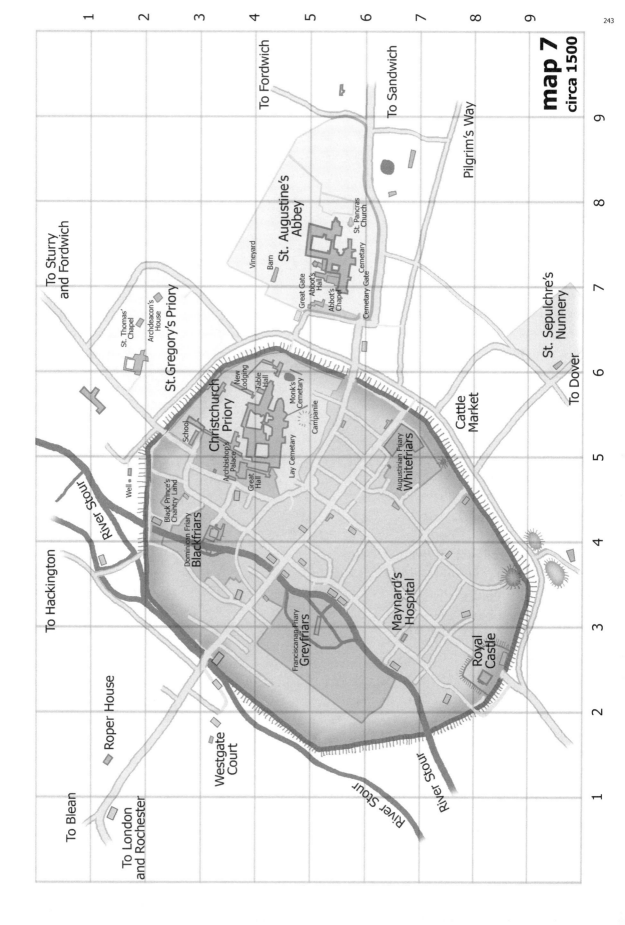

map 7
circa 1500

To Blean

To London
and Rochester

To Hackington

Roper House

To Sturry
and Fordwich

Westgate
Court

River Stour

River Stour

River Stour

Well

St. Thomas'
Chapel

Archdeacon's
House

St. Gregory's Priory

Black Prince's
Chantry Land

Dominican Friary
Blackfriars

School

Archbishop's
Palace

Great
Hall

Christchurch
Priory

New
Lodging

Table
Hall

Monk's
Cemetery

Campanile

Lay Cemetery

Franciscan Friary
Greyfriars

Maynard's
Hospital

Augustinian Friary
Whitefriars

Royal
Castle

Cattle
Market

To Fordwich

Vineyard

Barn

St. Augustine's
Abbey

Great Gate

Abbot's
Hall

Abbot's
Chapel

St. Pancras
Church

Cemetery

Cemetery Gate

To Sandwich

Pilgrim's Way

St. Sepulchre's
Nunnery

To Dover

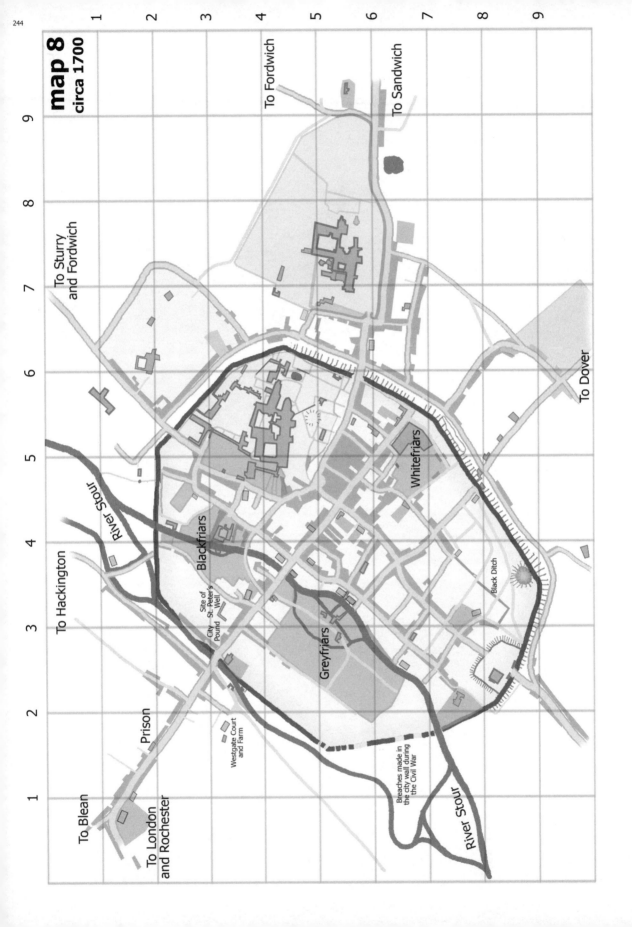

map 8
circa 1700

244

1 2 3 4 5 6 7 8 9

To Fordwich

To Sandwich

To Sturry
and Fordwich

To Dover

To Hackington

River Stour

To Blean

Prison

To London
and Rochester

River Stour

Westgate Court
and Farm

City
Pound

Site of
St. Peter's
Well

Blackfriars

Greyfriars

Whitefriars

Black Ditch

Breaches made in
the city wall during
the Civil War

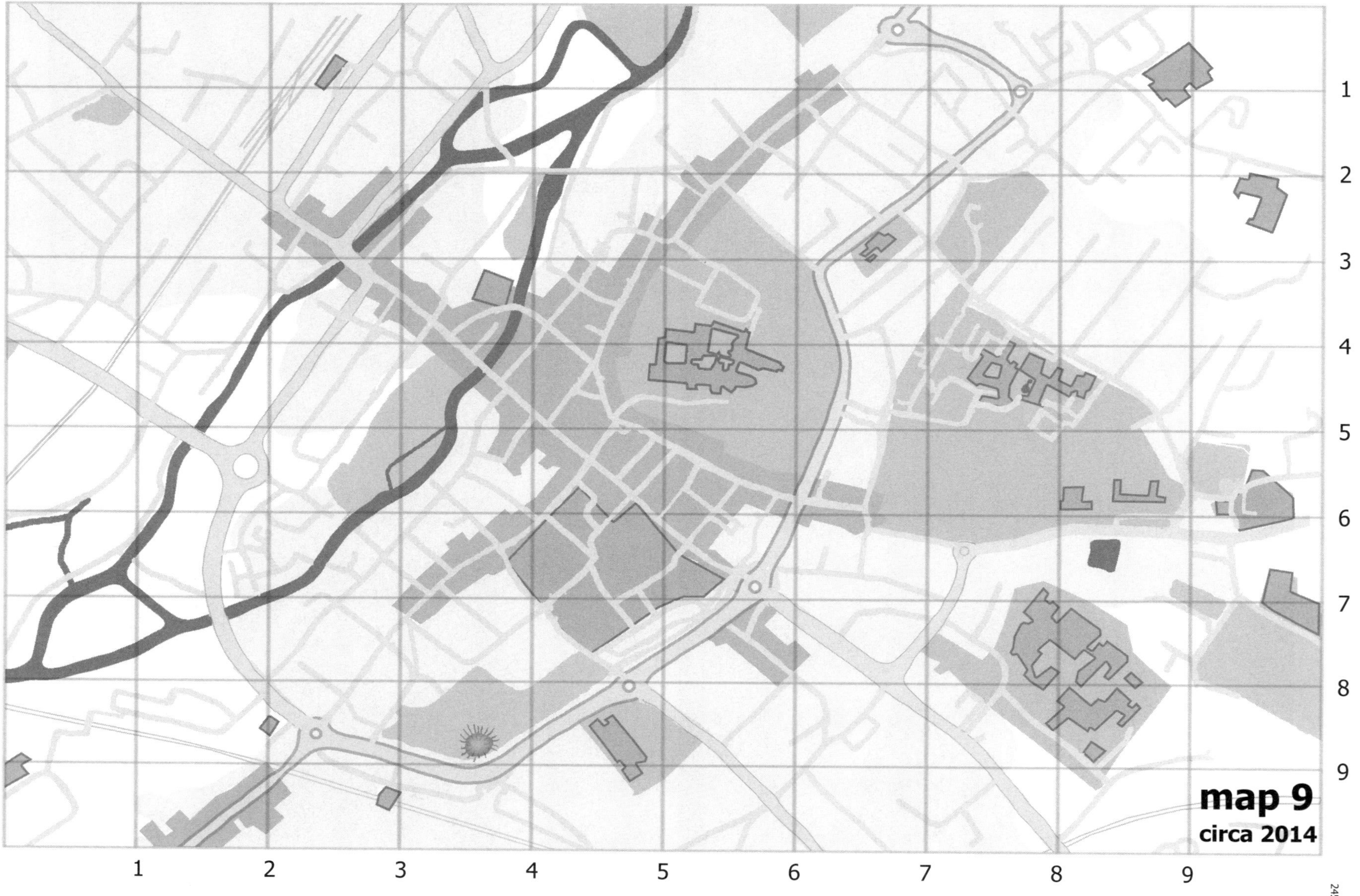

map 9
circa 2014

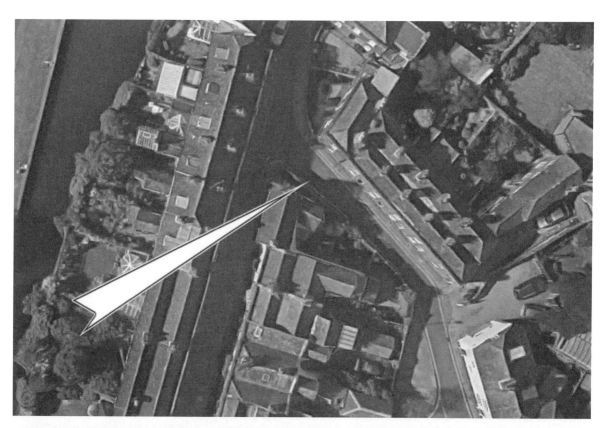

Appendix 1: key dates in the history of Canterbury

54BCE Julius Caesar briefly invades, does battle at a settlement (almost certainly Bigbury Hillfort)

43CE Roman invasion, city established

270-290 Roman city walls built

313 Constantine's Edict of Toleration: a Christian Church appears in the city soon after

410 Romans withdraw military protection of coast, city gradually abandoned

560 Ethelbert, married to Christian Frankish princess Bertha, becomes King of Kent

597 St. Augustine arrives with entourage from Rome, bringing Christianity

835-855 city suffers from regular Viking attacks

1011 Viking siege of the city, Cathedral burned, Archbishop Alphege kidnapped and ransomed

1067 rebuilt Cathedral completely burns down

1070 Lanfranc becomes the first Norman Archbishop of Canterbury

1085 Norman castle completed

1170 Archbishop Thomas Becket murdered in the Cathedral

1174 Henry II visits to atone for Becket's murder; another Cathedral fire

1188-89 Christchurch Priory besieged by Archbishop Baldwin and his supporters

1216 Castle under French occupation

1220 Becket's remains translated to a new shrine; Blackfriars arrive

1224 Greyfriars arrive and establish friary

1240 Cathedral monks excommunicated en masse by Archbishop Edmund

1261-1276 sporadic attacks on Jewish homes

1314 the radical Friars of the Sack banned and dispersed

1380 Westgate built, funded by Archbishop Sudbury

1381 Peasants' Revolt passes through city, marches up to London and beheads Sudbury

1382 a major earthquake brings down Cathedral bell tower

1415 Henry V welcomed in the city on his way home from the Battle of Agincourt

1448 a new charter replaces two bailiffs with an elected mayor

1450 Jack Cade's Rebellion

1461 city granted county status "forever" by Edward IV

1478 Geoffrey Chaucer's *The Canterbury Tales* appears in print

1500 Bell Harry, the Cathedral's central tower, now at its full height

1520 Henry VIII brings Emperor Charles V to the Jubilee for the translation of Becket's remains

1528 "Holy Maid of Kent" Elizabeth Barton becomes known to the public

1534 Elizabeth Barton taken to London to be hung

1538-9 Abbey and Cathedral Priory surrender, the three friaries suppressed

1555-58 forty-one Protestants burned by order of Queen Mary

1573 Elizabeth I visits for her 40th birthday

1575 first major influx of Huguenot immigrants

1626 Charles I marries Henrietta of France in the Cathedral

1636 William Somner publishes *The Antiquities of Canterbury*

1642 Colonel Sandys and company enter the city, find Royalist arms hidden in Cathedral, wreak havoc

1647 Christmas riots

1660 Charles II spends a night at St. Augustine's Abbey on his way back from exile

1703 a major storm badly damages the Cathedral's Arundel Tower

1717 *Kentish Post* first published

1730 Freemasons begin to operate in the city; Jews begin to resettle

1762 new synagogue opens in St. Dunstan's

1765 Mozart performs at the Guildhall

1766 George III grants a liberty making the city a centre of the hop trade

1768 *Kentish Gazette* first published, buys out the *Post* after a brief trade war.

1793 first Kent and Canterbury Hospital opens in Longport

1794 first military barracks established

1826 last Michaelmas Fair held in Cathedral Precincts

1830 Canterbury-Whitstable Railway opens

1832 Archbishop Howley's visit provokes a riot

1838 "Sir William Courtenay" dies at the so-called Battle of Bossenden Woods

1846 Canterbury West station and railway line to Ashford open

1859 Westgate very nearly destroyed for the sake of a visiting circus

1871 engineer James Pilbrow accidentally discovers Roman street layout

1891 George Beaney dies, leaving £10,000 to the city to create a library in his name

1910 The Electric Theatre, the city's first cinema, opens in St. Peter's Street

1935 T.S. Eliot's *Murder in the Cathedral* performed in the Cathedral

1936 Westgate Gardens presented to the people

1942 German "Baedekker" bombing raid on city causes untold damage

1952 Canterbury-Whitstable railway closes

1961 Australian beatnik Daevid Allen meets Simon Langton Boys' School pupil Robert Ellidge (Wyatt)

1962 Christchurch College (later CCCU) opens

1965 The Wilde Flowers form; the University of Kent at Canterbury opens

1966 The Soft Machine begin rehearsing in Sturry

1968 Caravan form, begin rehearsing in Whitstable; Caravan and Soft Machine record debut albums

1974 the city loses the county status granted to it in 1461, relocates Guildhall to Holy Cross church

1981 Odeon cinema becomes new Marlowe Theatre; Madonna and child statue stolen from Cathedral

1982 Pope John Paul II visits the city

1986 Thatcher and Mitterand sign "Treaty of Canterbury" in the Cathedral

1989 Cathedral, Abbey grounds and St. Martin's Church granted World Heritage status

1995 Cathedral begins charging entrance fee

2005 Whitefriars shopping development opens

2007 a leg of the Tour de France ends on Rheims Way

2008 Marlowe Theatre demolished to make way for a new one

2009 high-speed train link established with St. Pancras International station, London

2012 newly refurbished "Beaney House of Art & Knowledge" opens

2013 "5 Scots" regiment leaves the city, ending its history of hosting a military presence

2014 HMP St. Augustine's, the city's recently closed prison, is taken over by CCCU

Appendix 2: Archbishops of Canterbury

Saxon era
597 Augustine*
604 Laurentius*
619 Mellitus*
624 Justus*
627 Honorius*
655 Deusdedit*
664 *see vacant*
668 Theodore*
693 Berhtwald*
731 Tatwine*
735 Nothelm*
740 Cuthbert
761 Bregowine
765 Jaenberht
793 Aethelheard
805 Wulfred
832 Feologeld
833 Ceolnoth
870 Aethelred
890 Plegmund
914 Athelm
923 Wulfhelm
942 Oda
959 Brithelm
959 Aelfsige
960 Dunstan*
988 Aethelgar
990 Sigeric the Serious
995 Aelfric of Abingdon
1005 Alphege*
1013 Lyfing
1020 Aethelnoth
1038 Eadsige
1051 Robert of Jumièges
1052 Stigand

Norman era
1070 Lanfranc
1089 *see vacant*

1093 Anselm*
1109 *see vacant*
1114 Ralph d'Escures
1123 William de Corbeil
1136 *see vacant*
1139 Theobald
1161 see vacant
1162 Thomas Becket*
1174 Richard of Dover
1184 Baldwin of Forde
1191 *see vacant*
1193 Hubert Walter
1207 Stephen Langton
1229 Richard le Grant
1234 Edmund Rich*
1245 Boniface of Savoy
1273 Robert Kilwardby
1279 John Peckham
1294 Robert Winchelsey
1313 Walter Reynolds
1328 Simon Meopham
1333 John de Stratford
1349 Thomas Bradwardine
1349 Simon Islip
1366 Simon Langham
1368 William Whittlesey
1375 Simon Sudbury
1381 William Courtenay
1396 Thomas Arundel
1398 Roger Walden
1399 Thomas Arundel (restored)
1414 Henry Chichele
1443 John Stafford
1452 John Kempe
1454 Thomas Bourchier
1486 John Morton
1501 Henry Deane
1503 William Warham
1533 Thomas Cranmer
*recognised saint

post-Reformation era
1556 Reginald Pole
1559 Mathew Parker
1576 Edmund Grindal
1583 John Whitgift
1604 Richard Bancroft
1611 George Abbot
1633 William Laud
1645 *see vacant*

post-Commonwealth era
1660 William Juxon
1663 Gilbert Sheldon
1678 William Sancroft
1691 John Tillotson
1695 Thomas Tenison
1716 William Wake
1737 John Potter
1747 Thomas Herring
1757 Matthew Hutton
1758 Thomas Secker
1768 Frederick Cornwallis
1783 John Moore
1805 Charles Manners Sutton
1828 William Howley
1848 John Bird Sumner
1862 Charles Thomas Longley
1868 Archibald Campbell Tait
1883 Edward White Benson
1896 Frederick Temple
1903 Randall Thomas Davidson
1928 Cosmo Gordon Lang
1942 William Temple
1945 Geoffrey Francis Fisher
1961 Arthur Michael Ramsey
1974 Frederick Donald Coggan
1980 Robert Runcie
1991 George Carey
2002 Rowan Williams
2014 Justin Welby

Appendix 3: monarchs

Kentish monarchs
5th-6th century Oisc
c.512-c.534 Octa
c.534-c.590 Eormenric
c.590-616 Ethelbert I
616-640 Eadbald
(?) Ethelwald
640-664 Eorcenberht
(?) Eormenred
664-673 Ecgberht I
c.674-685 Hlothhere
c.685 to 686 Eadric
c.686-687 Mul
c.687-c.692 Swæfheard [*jointly*]
689 Swæfberht [*jointly*]
689-690 Oswine [*jointly*]
c.693-725 Wihtred [*jointly*]
725-? Alric [*jointly*]
725-748 Eadberht I [*jointly*]

Kent under Mercian overlordship
725-762 Ethelbert II [*jointly*]
(?) Eardwulf [*jointly*]
762 Eadberht II [*jointly*]
762 Sigered [*jointly*]
(?) Eanmund
764-765 Heaberht [*jointly*]
765-779 Ecgberht II [*jointly*]
784 Ealhmund
785-796 Offa of Mercia [*directly*]
796-798 Eadberht III Præn
807-809 Cuthred
809-? Cœnwulf
822-823 Ceolwulf
823-825 Baldred
825-839 Ecgberht III [*jointly*]
825-858 Ethelwulf [*jointly*]
839-851 Athelstan I [*jointly*]
855-866 Ethelbert III [*jointly*]
866-871 Ethelred I

English monarchs
871-899 Alfred the Great
899-924 Edward the Elder
924 Efweard
924-939 Athelstan
939-946 Edmund I
946-955 Eadred
955-959 Eadwig
959-975 Edgar the Peaceful
975-978 Edward the Martyr
979-1013 Ethelred the Unready
1013-1014 Sweyn Forkbeard
1014-1016 Ethelred the Unready
1016 Edmund Ironside
1016-1035 Cnut
1035-1040 Harold Harefoot
1040-1042 Harthacnut
1042-1066 Edward the Confessor
1066 Harold Godwinson
1066 Edgar the Etheling [*uncrowned*]
1066-1087 William I
1087-1100 William II
1100-1135 Henry I
1135-1154 Stephen
  [1141 Matilda]
1154-1189 Henry II
1189-1199 Richard I (Lionheart)
1199-1216 John
1216-1272 Henry III
1272-1307 Edward I
1307-1327 Edward II
1327-1377 Edward III
1377-1399 Richard II
1399-1413 Henry IV
1413-1422 Henry V
1422-1461 Henry VI
1461-1470 Edward IV
1470-1471 Henry VI
1471-1483 Edward IV
1483 Edward V [*uncrowned*]

1483-1485 Richard III
1485-1509 Henry VII
1509-1547 Henry VIII
1547-1553 Edward VI
1553 Jane [*uncrowned*]
1553-1558 Mary I
1558-1603 Elizabeth I
1603-1625 James I
1625-1649 Charles I
1649-1660 Commonwealth
1660-1685 Charles II
1685-1688 James II
1689-1694 William III and Mary II
1694-1702 William III
1702-1714 Anne

British monarchs
1714-1727 George I
1727-1760 George II
1760-1801 George III
1801-1820 George III
1820-1830 George IV
1830-1837 William IV
1837-1901 Victoria
1901-1910 Edward VII

United Kingdom monarchs
1910-1936 George V
1936 Edward VIII [*abdicated*]
1936-1952 George VI
1952- Elizabeth II

Appendix 4: recommended Canterbury discography

Daevid Allen Trio – *Live 1963* (released 1993)

The Wilde Flowers – *The Wilde Flowers* (compilation of demos 1965-67, released 2015)

Canterburied Sounds, volumes 1-4 (archival recordings compiled by Brian Hopper, 1998)

Soft Machine – *Jet-Propelled Photographs* (1967 demos, released 1977); ***The Soft Machine*** (1968); *Volume Two* (1969); ***Third*** (1970); *Fourth* (1971); *Fifth* (1972)

Caravan – *Caravan* (1968); ***If I Could Do It All Over Again***... (1970); ***In The Land Of Grey And Pink*** (1971)

Gong – ***Camembert Electrique*** (1971); *Flying Teapot* (1973); ***Angel's Egg*** (1973); *You* (1974)

Matching Mole – ***Matching Mole*** (1972); *Matching Mole's Little Red Record* (1972)

Hatfield and the North – ***Hatfield and the North*** (1974); *The Rotters' Club* (1975)

Robert Wyatt – *The End of an Ear* (1970); ***Rock Bottom*** (1974); *Ruth Is Stranger Than Richard* (1975)

Hugh Hopper – *1984* (1973); *Hopper Tunity Box* (1977)

Daevid Allen – *Banana Moon* (1971); *Good Morning* (1976)

Kevin Ayers – ***Joy of a Toy*** (1969); *Shooting At The Moon* (1970); *Whatevershebringswesing* (1971)

Steve Hillage – ***Fish Rising*** (1975); *L* (1976); *Green* (1978)

Spirogyra – *St. Radigunds* (1971); *Old Boot Wine* (1972); *Bells, Boots and Shambles* (1973)

This list only covers a small fraction of all the "Canterbury Scene" records out there. It partly reflects my tastes but I've also chosen to focus on bands and musicians with a direct link to the Canterbury area (hence the non-inclusion of Egg and National Health, widely considered "canonical" Canterbury Scene bands, and the inclusion of Spirogyra who are usually omitted from such lists). Titles in bold are my recommended starting points. To further explore this music I'd suggest listening to my *Canterbury Soundwaves* podcast series [http://canterburysoundwaves.blogspot.com]. Some of the Canterbury-based musicians and bands who are mentioned in the latter part of this book also have recordings available:

Syd Arthur – *Moving World* EP (2011); *On An On* (2012); *Sound Mirror* (2014); *Apricity* (2016)

The Boot Lagoon – *The Boot Lagoon* EP (2012)

Zoo For You – *Fast Dance On A Nail* (2012)

Koloto – *Mechanica* EP (2014)

Lapis Lazuli – *Extended Play* (2012); *Reality Is...* (2012); *Alien* (2014); *Wrong Meeting* (2016)

Arlet – *Arlet* EP (2012); *Clearing* (2013); *Quartet* EP (2014); *Big Red Sun* (2016); *Trio* EP (2016)

Kairo – *Kairo* EP (2014); Jamie Dams – *Rush of Souls* (2016)

Sam Bailey – *Free Range Piano* (2016); pianointhewoods.com

Delta Sleep – *Management* EP (2014)

Ed, Will & Ginger – *Songs* (2009); awalkaroundbritain.com

Bardo Thodol (most of whom lived at 12 Mill Lane) never made any studio recordings but a 2010 demo can be found online. Original members Jacob Brant and Anthony Saggers have both released extensive solo works of elegiac electronica and neo-classical compositions (as Yakobfinga and Stray Ghost, respectively).

For an eclectic mix of Canterbury sounds old and new, lend your ears to my *Canterbury Sans Frontières* podcast series [http://canterburywithoutborders.blogspot.com]. An audio archive of "Free Range" events is available at http://www.free-range.co (yes, that's ".co" – the site "lives" far from Canterbury, in Columbia).

Appendix 5: Yanik's poems (read on 14/08/14)

"Fire and the Dreamer"
In the valley, there's a fire, drying rivers with its breath
It does not fulfil itself, and the flames keep getting higher
In a carbon thirst, it bursts, darkening all human tracks
It reshapes the Earth and Moon, melts them into grey and black
Flames incinerating all
Like a thousand dreamless nights merge into a single dream
Forest waiting to ignite
In the valley, shaking, a dreamer dreams his own awakening
As the flames begin to tire,
thunder sounds its warning whisper
Now the dreamer is awake, and reality is fiction
It's reduced to ash that smoulders,
leaves are carried by the wind
and all the sparks race to caress them
Ancient woods thirst for ignition
The clouds rejoice in their red apparel
In the valley, there's a fire
And a dreamer, now awake
He does not fulfil himself, and the flames keep getting higher
With his last few dying breaths, to the sky all red and black,
he says, as follows,
"Ashes to ashes, dust to dust
Now I'm a wildfire,
rain if you must.

"The Gravity of the Situation"
Grave, lush stones of time found rest inside the chest,
infinity in half an hour
neurochemistry at its best
Chemicals, atoms and maggots
Beings made out of Teflon
Forever trying to stick,
or attempting to let go
From forever to nevermore
the heartbeat sounds just like terrorcore
Lifetimes carved out of errors
make for the best art, therefore,
I enjoy bliss but it's worthless,
I'll take wrong turns on purpose
I'll grasp for air and dive in hell
until I re-emerge wordless.

acknowledgements, notes and disclaimers

Many thanks to: Paul Davies for helpful layout suggestions; Paul Crampton, Brian Hopper and Angela Steddy for help on local historical details; Cathedral historians Cressida Williams and Madylene Outen, and whoever's behind the Cathedral's Twitter profile; Paul at SERCO who kindly talked me through the fate of purple bags of household waste collected in Canterbury in 2014; Dr. Don Aldiss, formerly of the British Geological Survey, for patient and thorough advice on accurately describing the formation of the Great Stour Valley; Greg Balco of the Berkeley Geochronology Center for advice on the British-Irish Ice Sheet; Kevin Twyman for providing family history and photos, and allowing those photos to be used on p. 207; Sarah Fillery for taking the cover photo; my lovely mother, Matt Tweed and Juliet Suzmeyan for all of their creative input, and in particular to Juliet for her cheerful willingness to enter into the spirit of this strange project of mine in which she found herself unexpectedly featuring.

I fully realise that my accounts of cosmology, geology, natural history and "prehistoric" human activities, may eventually come to be seen as completely wrong. In carrying out my research I realised the extent to which the further back you go in time, the sketchier and more speculative the available information becomes. I just had to work with the best data I could find at the time of writing and then use my best judgement.

I've chosen to use the (originally American, now fairly standard worldwide) version of "billlion" and "trillion", so one billion = 1,000,000,000 and one trillion = 1,000,000,000,000.

You may have noticed that both metric and imperial units are used, in a fairly haphazard mixture. This was intentional. For the pre-human phase of the book, metric units seemed appropriate, but while writing about the pre-metric human history of Canterbury, metres, kilograms, etc. somehow seemed inappropriate. This was also a nod to the (largely overlooked) value of preserving and studying ancient metrological systems.

Because locals have always referred to the entire stretch of road from the Westgate to (the site of) St. George's Gate as "the high street", I decided to use the lower case version to refer to all of St. Peter's Street, the High Street, Parade and St. George's Street, and the upper case version for just the stretch from The Friars to Mercery Lane which formally bears that name.

Tweets included as images look like they were posted at times which don't match the timeframes they appear in. This is simply because all tweets are timestamped according to the time in San Francisco where Twitter is based (an eight-hour difference with Canterbury time).

The neuroscientific commentary inset into the images on pp. 214-219 was put together under the careful guidance of Prof. Padmanabhan Sudevan (retired) of the University of Wisconsin. The timings of course cannot be absolutely certain but he assures me that they're certainly "plausible".

Living persons' full names have only been included if they were already in the public domain (via local politics, commerce, music, etc.), otherwise I've just used given names (and occasionally an initial to distinguish multiple persons with the same given name). If anyone appears in this book, photographically or otherwise, and would prefer not to, please get in touch via the website www.youareherebook.com and all future editions will be adjusted accordingly.

picture credits

cover photograph © 2016 Sarah Fillery; illustrations on pp. 2-38 © 2017 Matt Tweed; illustrations on pp. 39-41 © Carol J. Watkins and Matt Tweed; illustrations pp. 44-153 © 2017 Carol J. Watkins; illustrations on pp. 166-168, 198-220 © 2017 Juliet Suzmeyan; photographs not credited below © 2014 Matthew Watkins

p. 39: Juliberrie's Grave after William Stukeley; p. 41: after a display board in Canterbury Heritage Museum [artist unknown]; p. 45: Roman Canterbury c.150CE after a display board in Canterbury Roman Museum [artist unknown]; Roman Canterbury c.300CE after John A. Bowen; p. 50: Roman Ridingate and St. Martin's Church after John A. Bowen; p. 51: ruined amphitheatre after Ivan Lapper; 6th century resettlement after a display board in Canterbury Heritage Museum [artist unknown]; p. 58: motte-and-bailey after Laurie Sartin; stone castle after John A. Bowen; Saxon Cathedral after Ivan Lapper; p. 59: Lanfranc's Cathedral after John A. Bowen; p. 64: Abbey after Terry Ball; Westgate and Becket's shrine after Laurie Sartin; Blackfriars Gate after an 18th century drawing [artist unknown]; Blackfriars estate after an image from Canterbury Historical & Archaeological Society website [artist unknown]; p. 65: medieval Canterbury after John A. Bowen; Greyfriars estate after *Time Team* CGI reconstruction; p. 76: medieval pilgrims' inns after John A. Bowen; p. 77: Newingate and Northgate after period drawings [artists unknown]; Worthgate after William Stukeley; Burgate after John A. Bowen; p. 82: Buttermarket after a photo from the collection of Tina Machado; Christchurch Gate after Thomas Sidney Cooper; Dane John Gardens after George Shepherd; Ridingate after a period etching [artist unknown]; p. 83: All Saints Church after a period etching [artist unknown]; Wincheap Gate after S. Cooper; Catch Club after an 18th century etching [artist unknown]; p. 90: High Street coaching scene after Thomas Sidney Cooper; Electric Theatre after a photo from the collection of Arthur Lloyd; Curiosity Shoppe after a photo from the collection of Tina Machado; railway opening after drawing by Thomas Mann Baynes; p. 91: Battle of Bossenden Woods after an illustration from a 1900 book by John Campbell, 9th Duke of Argyll [artist unknown]; Thomas Sidney Cooper after Thomas Sidney Cooper; Ridingate after an 1882 photograph [photographer unknown]; St. George's Street after a photo from the *Kentish Gazette*; p. 96: St. Dunstan's Street after a photo from the collection of Paul Crampton; Rupert Bear in Fairyland after Mary Tourtel, from "Little Bear's Adventure", serialised in the *Daily Express*, 1921; Rupert Bear in tree after Alfred Bestall from "Rupert's Elfin Bell", serialised in the *Daily Express*, 1948; Simon Langton Grammar School after a photo from the collection of Paul Crampton; p. 97: WWI memorial unveiling after a photo from the collection of Tina Machado; William Temple after a photo by Hans Wild; bombing devastation of St. George's area, bomb damage to St. George's Church and bomb damage to Burgate after photos from the *Kentish Gazette*; Marlowe Theatre after a photo from the collection of Arthur Lloyd; p. 102: Daevid Allen and Soft Machine after photos by Mark Ellidge; Wilde Flowers after a photo from the *Kentish Gazette*; Caravan after a photo of unknown origin; Tanglewood after a photo by Leslie Hopper; p. 103: Kevin Ayers and Steve Hillage after a photo by Richard Imrie; Broad Oak festival roadblock after a photo from the collection of Alan Jones; UKC students arriving after a photo by Douglas Miller; collapse of Cornwallis Building after a photo from the collection of James B. Brown; Bagpuss and Clangers after photos from Smallfilms collection; p. 120: Mark Fuller's sculpture after a photo from the collection of Paul Crampton; p. 121: Post Office street art after a photo by Mark Jansen; p. 142: Crows v. Seagulls after an uncredited photo from Friends of Kingsmead Field website; city councillors after an uncredited photo from East Kent Liberal Democrats website; Lapis Lazuli after a photo by Maria Sullivan; Koloto after a photo by Bailladera; p. 143: Arlet after a photo by Alaric King; Syd Arthur after a photo by John Evans; Zoo For You after a photo (possibly) by Rosemary of Harbledown; Orange Street Music Club after photos by Sarah Yarwood; p. 152: Robert Wyatt after a photo by Matt Wilson; Rotary duck race after an uncredited photo from Canterbury Rotary Club website; anti-austerity march after an uncredited photo from KentOnline website; Davee Wilde after an uncredited photo from KentOnline website; p. 153: Hey Maggie after an uncredited promotional photo; Catching Lives *Monopoly* celebration after a photo by Brian Green; p. 155: rafting on the Stour after a photo by Avery McGuire; p. 205: from the collection of Tina Machado; p. 207: from the collection of Kevin Twyman; p. 209: Great Dunstan photo originally from *Kent Messenger*[?]; p. 211 Great Dunstan photo by Wikipedia user Poe123 [public domain]; pp. 238-242: maps 4-8 after maps drawn by Peter Bennett for the Canterbury Archaeological Trust

bibliography

The following books were all helpful in putting this one together and contain a wealth of further information and imagery for those wishing to learn more:

John Brent, *Canterbury in the Olden Time* (Ginder, 1860)

Audrey Bateman, *Hail Mother of England!: a social history of Canterbury* (Rochester Press, 1984)

Ivan Green, *Canterbury: a pictorial history* (Phillimore & Co, 1998)

Dominic Bellenger and Stella Fletcher, *The Mitre and the Crown: a history of the Archbishops of Canterbury* (Sutton, 2005)

Graham Bennett, *Soft Machine: Out-bloody-rageous* (SAF, 2005)

Marjorie Lyle, *Canterbury: 2000 years of history* (Tempus, 2008)

Paul Crampton, *The Canterbury Book of Days* (History Press, 2012)

Jens Harder, *Alpha: Directions* (Knockabout, 2015)

David Birmingham, *Canterbury Before the Normans* (Palatine Books, 2016)

These also informed the later parts of the narrative, albeit in a less direct and "factual" way:

Hilaire Belloc, *The Old Road: from Canterbury to Winchester* (Constable, 1904)

Kahlil Gibran, "The Three Ants", from *The Madman* (Knopf, 1918)

James Joyce, *Ulysses* (Sylvia Beach, 1922)

Ludwig Wittgenstein, §164 from *Philosophical Investigations* (Blackwell, 1953)

Jorge Luis Borges, "The Aleph", from *The Aleph and Other Stories* (Jonathan Cape, 1971)

Russell Hoban, *Riddley Walker* (Jonathan Cape, 1980)